SOIL
TESTING
for Engineers

SERIES IN SOIL ENGINEERING

Edited by

T. William Lambe
Robert V. Whitman
Professors of Civil Engineering
Massachusetts Institute of Technology

BOOKS IN SERIES:

Soil Testing for Engineers by T. William Lambe, 1951
Soil Mechanics by T. William Lambe and Robert V. Whitman, 1969
Soil Dynamics by Robert V. Whitman (*in progress*)
Fundamentals of Soil Behavior by James K. Mitchell (*in progress*)

The aim of this series is to present the modern concepts of soil engineering, which is the science and technology of soils and their application to problems in civil engineering. The word "soil" is interpreted broadly to include all earth materials whose properties and behavior influence civil engineering construction.

Soil engineering is founded upon many basic disciplines: mechanics and dynamics; physical geology and engineering geology; clay mineralogy and colloidal chemistry; and mechanics of granular systems and fluid mechanics. Principles from these basic disciplines are backed by experimental evidence from laboratory and field investigations and from observations on actual structures. Judgment derived from experience and engineering economics are central to soil engineering.

The books in this series are intended primarily for use in university courses, at both the undergraduate and graduate levels. The editors also expect that all of the books will serve as valuable reference material for practicing engineers.

T. William Lambe and Robert V. Whitman

SOIL
TESTING
for Engineers

T. WILLIAM LAMBE
The Massachusetts Institute of Technology

NEW YORK · JOHN WILEY & SONS, INC.
LONDON · SYDNEY

ISBN 0 471 51183 8

PRINTED IN THE UNITED STATES OF AMERICA

Preface

The branch of engineering known as soil mechanics is still relatively young. Since its initiation, the number of colleges and universities giving formal courses in soils has steadily increased until today nearly every engineering school offers at least one course in soil mechanics. In the past, many instructors taught from their personal notes for lack of adequate textbooks. In recent years, however, several good books on the theory and practice of soil mechanics have appeared. The primary purpose of this book is to fulfill the need for a textbook for the teaching of soil testing. I think that it will also be of value as a reference book for practicing engineers and for personnel in soil laboratories.

There are many conditions, such as soil disturbance during sampling and unknown boundary conditions in nature, that require the engineer to use intelligence and experience in applying the results of laboratory tests to an actual soil problem. In many particular problems the results of laboratory tests can do no better than serve as a guide to the designer. For such problems field tests are often required. Even though this book is devoted entirely to laboratory testing, I hope that it will aid the student in attaining the proper perspective on the role of laboratory testing in soil mechanics.

In addition to an introductory chapter on general laboratory procedures, this book devotes a chapter to each of the tests presented. Only the common laboratory soil tests are included; several tests which are of a semi-empirical nature and the results of which cannot be well interpreted have been excluded. Also specialized research tests have not been included herein. Consideration was given to the inclusion of some clay technology and the related laboratory tests, such as the determination of clay minerals, nature and amount of exchangeable ions, electrical charges on particles, and degree of acidity (pH). Although rapid progress is being made in the development of clay technology, I feel that not enough is known concerning the influence of these clay properties on the engineering behavior of soil to justify their inclusion in this book.

Each chapter, after the first one, consists of sections entitled: Introduction, Apparatus and Supplies, Recommended Procedure, Discussion of Procedure, Calculations, Results, and Numerical Example. In the Introduction the property which is sought is defined and discussed, and practical applications of it are given. Under Apparatus and Supplies a list of the necessary equipment is given, along with photographs,* sketches, and any description thought desirable to illustrate important features. The reader should note that the detailed test procedures presented in this textbook are entitled Recommended Procedures. The exact procedure and type of apparatus which should be employed in any particular test depend on the soil in question and the use to be made of the results. An attempt has been made to give the principles of the laboratory

---

* Volume VI of *International Soil Mechanics Conference Proceedings*, Rotterdam, June, 1948, contains photographs and descriptions of the testing apparatus used by most of the major soil laboratories in the world. Unless otherwise noted, all photographs in this book were taken in the soil mechanics laboratory at the Massachusetts Institute of Technology.

tests in such a way that an instructor or a trained technician can see where he may alter the procedure to fit his particular needs. This book would perform a disservice to the profession if it posed as a manual and implied that a reader can always blindly follow detailed test procedures to obtain results which have much value. The results of soil tests should be interpreted by someone who understands the basic fundamentals.

As an illustration of the detailed computations, a numerical example is included at the end of each test. These examples employ recommended data-calculation sheets which minimize computation time and mistakes. A laboratory will find it advisable to stock a supply of data-calculation forms.

In an attempt to make this book reasonably self-sufficient, brief derivations of the formulas used in the test computations are included in the Appendix. Also in the Appendix are presented explanations and discussions of calibration procedures and special techniques, and other useful information.

I am indebted to the people who permitted me to use their work in this book. A concerted effort has been made to give proper credit to them for their contribution at the place where it was used. I would like to express my appreciation to my colleagues in soil mechanics at the Massachusetts Institute of Technology, Harl P. Aldrich, Paul B. Lawrence, C. C. Lu, W. R. Sutherland, and especially Professor D. W. Taylor, for reviewing the manuscript and making helpful suggestions. Miss Winifred Cadbury deserves credit for her aid in making this book more readable. Miss Dale Davisson and Miss Elizabeth Lilly typed and reviewed the manuscript. To Dr. Victor F. B. de Mello, a former member of the soil mechanics staff at the Massachusetts Institute of Technology, especial thanks are due for his sharp but constructive criticism based on a careful study of the manuscript.

T. William Lambe

Massachusetts Institute of Technology
June, 1951

Contents

Nomenclature

The nomenclature listed here agrees essentially with that suggested in the *American Society of Civil Engineers Manual 22,* entitled "Soil Mechanics Nomenclature."

A = area
A_0 = initial area
$\mathring{A}$ = angstrom units
a = area
a_v = coefficient of compressibility
B = water-plasticity ratio
b = depth
C_c = compression index
c_v = coefficient of consolidation
D = diameter
D_0 = initial diameter
d = diameter
d_f = final compression dial reading in consolidation test
d_0 = compression dial reading at zero time in consolidation test
d_s = corrected compression dial reading at zero time in consolidation test
d_{90} = compression dial reading at 90% primary compression in consolidation test
d_{100} = compression dial reading at 100% primary compression in consolidation test
e = void ratio
e_c = critical void ratio
e_0 = initial void ratio
F = force
f_c = compressive strength
G = specific gravity
G_s = specific gravity of soil particles
G_T = specific gravity of water at temperature T
H = height of soil sample
H_0 = height of soil specimen when the void volume is zero
h = head of water
h_c = capillary head
h_c' = effective capillary head
h_{cs} = saturation capillary head
h_0 = applied water head
I_f = flow index
I_p = plasticity index
I_t = toughness index
k = permeability
k_s = permeability at degree of saturation S
L = length
L_0 = initial length
$m = \dfrac{\Delta(x^2)}{\Delta t}$ in horizontal capillary flow
N = percentage by weight finer than any given diameter
N' = corrected percentage by weight finer than any given diameter in a combined grain size analysis
n = porosity
P = applied force
p = applied stress
Q = quantity of flow
Q = unconsolidated-undrained shear test
Q_c = consolidated-undrained shear test
q = rate of flow

R = 1000 (specific gravity hydrometer reading $-$ 1); radius
R_w = 1000 (specific gravity hydrometer reading in water $-$ 1)
r = specific gravity hydrometer reading; primary compression ratio
r_w = specific gravity hydrometer reading in water
S = degree of saturation
S = drained shear test
s = shear strength
T = temperature
T_s = surface tension
t = time
t_{90} = time at 90% consolidation
U = unconfined compression test
V = volume
V_s = volume of soil grains
V_v = volume of voids
V_w = volume of water
W = weight
W_s = weight of dry soil grains
W_w = weight of pore water
w = water content
w_l = liquid limit
w_n = natural water content
w_p = plastic limit
w_s = shrinkage limit
y = coordinate distance
Z_r = vertical distance from surface of a suspension to the center of volume of an immersed hydrometer
Z = coordinate distance
α = used for various angles
β = used for various angles
γ_a = unit weight of air
γ_d = dry density
γ_w = unit weight of water
γ_s = unit weight of dry soil grains
ϵ = thermal coefficient of cubical expansion for glass; strain
θ = angle between major principal plane and failure plane
μ = viscosity; microns
σ = stress perpendicular to the surface to which it is applied
$\bar{\sigma}$ = intergranular stress perpendicular to the surface to which it is applied
$\bar{\sigma}_1$ = major principal intergranular stress
$\bar{\sigma}_2$ = intermediate principal intergranular stress
$\bar{\sigma}_3$ = minor principal intergranular stress
$\bar{\sigma}_c$ = maximum pressure to which a clay has been consolidated
σ_{ch} = chamber pressure in triaxial compression test
σ_{ff} = the stress perpendicular to the failure surface at failure
τ = shear stress
τ_{cr} = shear stress on plane of maximum obliquity
ϕ = angle of internal friction
ϕ_a = apparent angle of internal friction
ϕ_m = peak angle of internal friction
ϕ_u = ultimate angle of internal friction

CHAPTER

I

Introduction

The answer to a problem in soil engineering is normally obtained by first determining the properties of the soil in question and then employing these properties to work out the solution. Since the soil at every site is different, the soil involved in each different problem must be evaluated. Often this evaluation can be approximated from a knowledge of the geology of the site or from experience with similar soils. Usually, however, the soil properties must be determined by laboratory or field tests.

This book deals with the laboratory testing of soils. In order to give the reader some help in acquiring an overall picture, examples are given of the types of problems which involve the soil property under discussion. For each test a brief treatment of the fundamentals involved is given in order that the principles of the test may be better understood. The intention herein is to present information which will serve as a guide for the determination of soil properties to suit particular boundary conditions. The fact that field conditions are not always completely known, or that methods of analysis may be inaccurate, is not considered justification for obtaining soil properties which cannot be interpreted.

Beginning with Chapter II, a complete chapter has been allotted to each laboratory test covered. There are several observations, determinations, and techniques, however, which are required in more than one test; these are presented in the first chapter.

Soil Identification and Description

Before testing a soil the technician should examine his sample; he should then note on his data sheet a general classification and should give a careful description of the soil and its condition. There are

three reasons why this procedure is extremely desirable. First, the procedure materially aids the development of a feel for soil behavior; second, it may aid in the interpretation of the results obtained; and, third, a check of the laboratory classification against that made in the field and recorded on the log of boring may detect a possible error in the sample numbering. As a technician tests a soil which he has previously attempted to classify and describe, he will unconsciously begin to relate his description and the test behavior of the soil. Every bit of such related data goes to build valuable judgment and experience. An experienced technician can often predict with surprising accuracy the behavior of a soil by working it in his hand and inspecting it carefully.

The noting in the sample description of such items as the presence of shells, varves, roots, cracks and prominent waves, and disturbance in the soil structure, may aid in the explanation of any unusual test results obtained. If the technician carefully observes and appraises his specimen before testing, he is more likely to notice any mistake such as a misplaced decimal point. At the conclusion of a test he should check to see if his results are in agreement with what he would expect from his classification.

There are many detailed soil classifications [1] currently employed. Most of these classifications were designed for use in a particular branch of soil work and are, therefore, limited in their application. For example, because the United States Bureau of Public Roads classification was developed for use with soils in road construction, it is not very helpful to the

[1] See reference I–4 for a description of the better-known soil classifications.

1

building foundation engineer. The soil classification [2] in this book is a general one which employs a soil description. Because this classification is broader and less arbitrary than the others, it is recommended for student use.

In classification, the soil in question is first relegated into one of the following groups: boulders, gravel, sand, peat, silt, or clay. Boulders are particles larger than 6 to 8 in. in diameter. Gravel consists of particles varying in size [3] from $\frac{1}{4}$ in. to 6 to 8 in. Sand sizes range from the lower limit of gravel to about 0.06 mm, which is approximately the size of the smallest particle that can be seen with the unaided eye. Peat consists largely of undecomposed and partly decomposed vegetable matter; twigs, roots, or leaves are usually visible in peat. The differentiation between a clay and a silt [4] can be based on the presence or lack of plasticity, dilatancy, and dry strength better than on grain size.

A measure of the plasticity of a soil can be obtained by noting how much working of the soil between the fingers is required to dry it from a wet state, near the liquid limit, to the crumbling state, near the plastic limit (see Chapter III). The greater the plasticity, the longer the kneading time required. This is true for two reasons: first, the more plastic the soil, the larger the difference in water contents between the wet and crumbling stages; second, the more plastic the soil, the more the effort required to expel a given quantity of water. Because silts possess little plasticity they usually dry after a few minutes of working with the fingers.

The dilatancy test (also known as water mobility or shaking test) consists of placing a pat of moist soil in the palm of the hand, and then shaking the hand. If a shiny, moist surface appears on the soil after shaking it in the open hand and then becomes dull and dry when the pat is squeezed by closing the hand, a nonplastic soil (e.g., silt) is indicated. Squeezing the pat causes shearing deformations in the soil, which in turn cause a nonplastic soil to expand from the low volume obtained by shaking. [5] The water flows into the soil to make up for the increased volume of pores and, thereby, leaves the surface with a dry appearance.

A dry piece of clay normally possesses appreciable dry strength, whereas a dry lump of silt can be easily powdered with the fingers. Thus a good test is to dry a small lump of the soil in question (the specimen should be dried slowly to avoid cracking) and obtain an indication of its dry strength by trying to powder the lump between the fingers.

After a soil has been identified as one of the previously described groups, or a mixture of them, it should be described. [6] A description should include items which may be helpful in predicting the behavior of the soil as well as those which help characterize it. For describing any soil the following may be considered.

1. *Color.* Since the color of a soil is dependent on its water content, the water content at which the color was observed (e.g., natural water content, air dry, etc.) should be noted if it is not otherwise clear. In addition to its value in characterizing a soil, color can be useful in indicating the presence of such items as organic matter and iron.

2. *Odor.* Occasionally soils have distinctive odors; however, a soil can lose its original odor or obtain a different one by adsorption while it is being stored in the laboratory.

3. *Minerals.* The behavior of a fine-grained soil is considerably influenced by the minerals [7] which make up the soil; unfortunately, the mineral components of such soils are not readily identified. On the other hand, the behavior of a coarse-grained soil is normally only slightly influenced by its minerals; these minerals can usually be readily identified.

4. *Presence of organic matter.* Organic matter can usually be detected by its characteristic odor and dark color. Organic matter tends to make a soil weaker and more compressible.

5. *Presence of foreign matter.* Such items as shells or rubbish often give a key to the origin of a soil.

[2] The Airfield Classification (I–4) developed by A. Casagrande has gained wide acceptance in many branches of soil engineering. It is also the basis of other classifications. (For example, the classification used by the Bureau of Reclamation. See reference I–1.)

[3] Two mm is also used as the lower range of gravel size.

[4] The distinction between silt and clay can be based on either grain size or behavior. For example, the M.I.T. grain size classification (see Fig. IV–8) gives 0.002 mm as the border between silt and clay. As will be seen in Chapter IV, the determination of percentage of a soil finer than this size is a time-consuming operation. Also it is more logical to classify a fine-grained soil according to behavior. In Chapter III is a method of distinguishing between silt and clay on the basis of plasticity. Because plasticity depends on particle size in addition to particle shape and composition, the two criteria are somewhat related. Behavior is the criterion for the differentiation described above. To avoid confusion, silt size and clay size are often employed when the distinction is based on size rather than behavior.

[5] The volume changes associated with shear deformations are discussed in Chapter X.

[6] A hand lens often materially aids the examination of a soil.

[7] The minerals which compose the clays generally belong to one of the following groups: montmorillonite, illite, and kaolinite.

6. *Geological history.* Sometimes the origin of a soil can be detected by examination. For example, varved clay (see Fig. I–3) can be recognized.

In describing gravels and sands, we should record degree of uniformity (see Chapter IV) and particle shape. For silts and clays the degree of plasticity, dry strength, dilatancy, "sensitivity" (see Chapter XII), consistency, homogeneity, as well as the presence of varves, cracks, or signs of disturbance, may be recorded. Any local names for a soil should be given.

A student taking his first laboratory course in soil mechanics should not be expected to classify and describe completely the soil used in the early tests, since several of these very tests are "classifying tests." However, from the feel obtained by observing the soil in the laboratory and from the experience gained from the classifying tests, a student should soon learn how properly to classify and describe the more common types of soil.

The Storage of Soil Samples

To obtain either undisturbed soil samples or disturbed samples which are representative of a soil stratum involves a process which may require skill, knowledge, and experience. The art of sampling has received much study, and a monumental publication (I–7) on it has been prepared by Hvorslev. To use highly perfected laboratory tests on soil samples which were improperly obtained appears to be inconsistent, because the results can be no better than the samples tested.[8] However, since sampling and related subjects, such as the shipment of samples, are outside the scope of this book, they will not be presented here.

Soil samples should be inspected and tested shortly after their arrival at the laboratory because of the difficulty of storing them properly and because the results of tests on them may indicate the desirability of altering the planned program of sampling in the field. Although immediate testing is desirable, many laboratories find sample storage a necessity. Storage may be needed if the testing cannot proceed as rapidly as the sampling or if samples are required for future research.

Adequate space is the main requisite for the storage of soils if the prevention of moisture change is not a requirement. Bags of canvas or unbleached and tightly woven duck, as well as cans and bins, are used for large quantities of soil; glass jars are satisfactory for small quantities. The container should have a label or tag which gives such data as soil type, loca-

tion of site where sample was taken, sampling date, boring number, sample depth, and sample number. If a file on the soil sample is kept, the label or tag should list the file number. An example of a label for samples and an example of a companion form for efficient recording of information on the sample are shown in Fig. I–1.

Undisturbed clay samples are generally of one of two types: pit samples or samples taken by some sort of sampler. Because pit samples, or those cut from an exposed surface of soil (I–14), can be the source of the least disturbed clay, they are taken, if feasible. After extraction, the sample is covered with a protective coating; this coating is applied either in the field or in the laboratory, depending on such considerations as the proximity of the site to the laboratory.

Although paraffin is commonly used as a coating, Osterberg's work (I–10) shows paraffin to be the poorest of the thirteen waxes which he investigated. The waxes which performed the best were:

Wax	*Melting Point*	*Manufacturer*
Petrowax A	168° F	Gulf Oil Corporation
50% Petrolatum + 50% paraffin	120°–125° F	Standard Oil Company of Indiana
Product 2300	155° F	Socony Vacuum Oil Company

Not only were the above waxes less brittle and less susceptible to shrinkage on cooling than the paraffin alone, but they were also considerably more impermeable to water. A moist clay sample protected with paraffin lost more than 270 times as much weight (approximately 8% of total weight) in a period of 300 days as did similar samples covered with the three waxes listed above and exposed under similar conditions. Petrowax A cracked much less than the other two waxes when subjected to temperatures below freezing.

The wax coating can be applied by either dipping the soil sample in the melted wax or using a soft brush to spread the melted wax. A wax should not be heated to more than a few degrees above its melting temperature, since heating to higher temperatures tends to drive off some of the more volatile hydrocarbons, thus making the wax more permeable and more brittle upon cooling.

The shrinkage of a wax can be decreased and the strength increased by including one layer, or more, of cheesecloth between wax coats. If the soil sample is to be stored for more than a few days, a total thickness of protective coating of ½ to ¾ in. is desirable. Such large thicknesses can be easily obtained by plac-

---

[8] The reader is referred to reference I–12 for a discussion of the relation of undisturbed sampling to laboratory testing.

ing the sample in a mold which is more than ½ in. larger than the sample, and then filling in the gap with melted wax.

Samples taken with a sampler may be stored in the tube in which they were extracted from the ground, or removed from the tube and stored in the same manner as pit samples. If a sampler, such as the fixed-piston type, is employed which furnishes a sample in a thin-walled tube, the storage of the sample in this tube is convenient. Wax is commonly poured into the cleaned ends of the tube to form a plug about an inch thick. A metal or rubber cap placed over the wax may help retard its plastic flow.

Soil kept in metal tubes may be seriously altered by electrolytic or chemical action. Discoloration and drying have been noted in clays stored in steel tubes for only short periods of time. Such action may be slowed down, or possibly prevented, if the inside of the tube is carefully lacquered or oiled.

Undisturbed samples, like the disturbed ones, should be carefully tagged or labeled. Paint, wax pencil, etc., can be used to mark on a tube; labels written with wax pencil, waterproof ink, etc., can be stuck to the sample with a coat of wax or lacquer. Tags can be attached to a tube by means of a wire looped through a hole drilled at the end of the tube. In addition to giving the data listed on page 3, one should carefully mark the top and bottom of an undisturbed sample. The label shown in Fig. I–1*a* is marked "Top of Sample" and, therefore, should be attached to the top.

The tendency of a wax to permit escape of soil water or to undergo plastic flow can be reduced by storing soil samples in a cool room in which the relative humidity is kept near 100%. Such a room is usually called a humid room. The stacking of samples on top of each other is not desirable since plastic flow is accelerated by stress. Storing tube samples

TOP OF SAMPLE

PROJECT__________ FILE NO.__________
LOCATION__________ DATE__________
__________ SAMPLE NO.__________
BORING NO.__________
DESCRIPTION OF SOIL__________
DEPTH TO SAMPLE TOP__________ BOTTOM__________
ELEVATION OF GROUND SURFACE__________
REMARKS__________

SAMPLE LABEL

(*a*)

PROJECT__________ FILE NO.__________
LOCATION__________ DATE__________
__________ SAMPLE NO.__________
BORING NO.__________
DESCRIPTION OF SOIL__________
DEPTH TO SAMPLE TOP__________ BOTTOM__________
ELEVATION OF GROUND SURFACE__________
NUMBER OF SAMPLES IN SHIPMENT__________
LOCATION OF SAMPLE IN LABORATORY__________
TYPE OF SAMPLE (pit, tube, jar, etc.)__________
ORIGINAL SIZE OF SAMPLE__________
CONDITION OF SAMPLE__________
REMARKS (tests run, reference to where test data on sample may be found, etc.)

SIZE OF SAMPLE STORED AFTER PROJECT COMPLETED__________

SAMPLE DATA FOR FILE

(*b*)

Figure I–1

vertically tends to prevent the formation of air channels by plastic flow of the wax.

The Handling of Undisturbed Samples

The serious effects that disturbance and loss of moisture can have on undisturbed cohesive soils are emphasized in the chapters on the consolidation and shear tests. Because of these effects, every reasonable precaution should be taken in the handling of undisturbed samples. Some of these precautions, in addition to trimming techniques, are described below.

Whenever and wherever possible, all preparation of undisturbed test specimens should be done in a humid room. A technician should never handle an unprotected sample with bare hands because the hands foster disturbance and loss of moisture. He can protect the sample by using something like cellophane or wax paper between his hands and the soil. When transporting a specimen, he should support it over its entire length by using a mold, plate, or paper sling.

In the following chapters the trimming of the particular test specimen required is described in the recommended procedures; these trimming instructions assume the possession of a chunk of soil slightly larger than the finished specimen. As stated on page 3, undisturbed clay samples are generally obtained either by cutting from a pit or by a sampler. The cutting of a chunk from a pit sample can be done with a wire saw which consists of a frame with a piano wire strung tightly across it. Wire saws are shown in Figs. IX–7 and XII–6. Any wax or other covering used to protect the sample should be cut with a knife or hack saw because a wire will not cut a thick covering easily and may tear it and adjacent soil. Some sort of guide, such as a miter box (see Fig. XII–6a), can be used to obtain a straight cut.

More care should be exercised when trimming the finished specimen. The wire in a wire saw used for final trimming is usually smaller in diameter than that used for the preliminary cutting. The wire should be cleaned prior to each cut because a piece of soil sticking to it may dig a groove in the finished surface. Certain soils, such as those that have a consistency like that of yeast, can be more easily cut with a sharp knife than with a wire saw which tends to tear. In fact, the cutting of most soils with a sharp knife results in less soil disturbance than cutting with a wire. On a surface cut with wire a thin layer of disturbed soil can often be detected from the smearing of stratification layers. No such smearing is visible on a surface cut with a sharp knife. In Fig. XIII–2 is shown a blade made from a straight razor.

The cutting of specimens from samples taken with a sampler is similar to the cutting of specimens from pit samples after the soil has been removed from the tube. Samples should be extruded from their tubes by a steady pushing process and not by a jerky or driving one. Prior to extrusion, however, the plug which protects the end of the soil should be removed by some means, such as scraping it out with a knife or sawing off that portion of the tube which contains the plug, and any burrs on the inside of the tube should be carefully eliminated with a scraper or file. The extrusion should employ the same direction of motion of the soil with respect to the tube as existed during sampling, because a reversal of stress tends to cause disturbance.

In Fig. I–2 is shown a satisfactory sample extruder. Rotation of the crank (Fig. I–2a) applies a force to the sample through the piston attached to the rack and pinion unit. The rack and pinion setup permits the application of a steady force to the sample. In Fig. I–2b is shown the detachable trough connected to the end of the extruder. Wax paper is fed between the trough and the extruder as the sample comes out so that the sliding is between the paper and the trough. The trough supports the sample over its entire length; later, it can be easily detached and used to transport the sample. The extruder can be adapted to handle tubes of other sizes by merely replacing the sliding plate at the end with one which has an opening of the size desired. By this means the extruder can be converted to the proper size to extrude a compaction sample (see Chapter V) from its mold.

The problem of sample removal can be avoided by testing the soil in its sample tube. For example, permeability tests can often be run on specimens in their sample tubes. In addition, split samplers lined with thin rings are in use (I–6). A technician can open the split sampler and remove the rings filled with soil; if the apparatus is designed for samples the size of the rings, he can easily cut off a ring of soil for testing. The use of samplers lined with rings greatly simplifies the preparation of test specimens, insures better fitting between the soil and test apparatus, and reduces soil disturbances from specimen trimming. Such samplers, however, are, of necessity, thick-walled, and, therefore, do not obtain as undisturbed a sample as thin-walled samplers.

Partially Dried Clay Specimens

To study a clay specimen in detail for any possible distortion obtained during sampling or for the presence or absence of homogeneity is often desirable. Heterogeneous clay specimens will often appear al-

(*a*)

(*b*)

FIGURE I–2. Sample extruder.

most homogeneous when they are at their natural water content or in a dried state, but in a partially dried state they will exhibit their heterogeneous nature to a marked degree. If a stratified soil, such as the normal sedimentary clay, is allowed to dry slowly, the coarser-grained strata will lose their pore water more rapidly than the finer-grained ones. As the water leaves the coarser grains, they become much

cracked are composed of clay; the others are of silt. The clay possessed more water in its natural state and thus lost more water during drying than the silt.

Partially dried specimens, such as the one in Fig. I–3b, offer a means of obtaining a geological record of a boring; of constructing better geological profiles; of selecting specimens for laboratory testing; of evaluating the quality of sampling; and of recording the na-

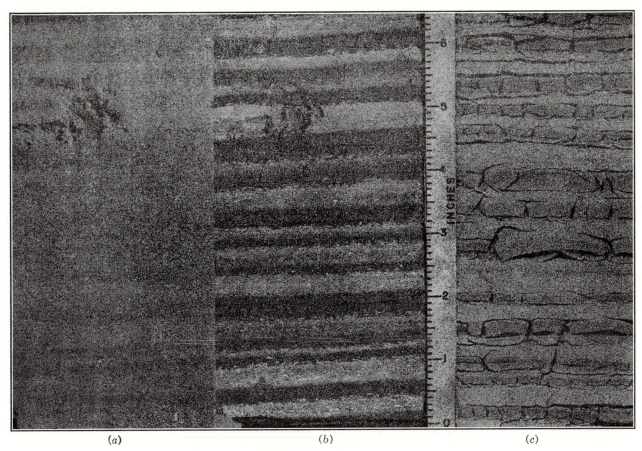

(a)   (b)   (c)

FIGURE I–3. Canadian varved clay at various stages of drying.

lighter in color and furnish a sharp contrast to the dark-colored portions.

Figure I–3 shows a Canadian varved clay at three different water contents. The specimen [9] in Fig. I–3a is at its natural water content; the specimen in Fig. I–3c is air dry; and the specimen in Fig. I–3b is at a water content between the other two. Although the stratification layers can be detected in Figs. I–3a and I–3c, they show much more clearly in Fig. I–3b. Soils which are more nearly uniform in grain size, such as the typical sedimentary clays, usually barely show the stratifications when at natural water content or when dry. The layers in Fig. I–3c which shrank and

ture of a tested specimen (see Fig. I–5b). A permanent record of a partially dried specimen can best be obtained by photographing the specimen at the proper stage of dryness. The method of preparing a specimen for photographing is presented briefly below; a much more complete presentation can be found in reference I–7.

In general the procedure of preparing partially dried specimens consists of three steps: slicing, drying, and trimming. The exact method of slicing depends somewhat on the type of specimen being dried as well as the use to be made of it. For example, if an entire sample is to be dried, no slicing is necessary (see Fig. I–5b). On the other hand, if studies are being made to select specimens for testing, longitudinal slices

[9] The same specimen was used for Figs. I–3a and I–3b; the specimen in Fig. I–3c is of an adjacent piece of the sample.

from the sample are required (see Fig. I–5a). Figure I–4 shows the steps in slicing a sample as worked out by Hvorslev (I–7). If the sample to be sliced is large, i.e., above 4 or 5 in. in diameter, the use of a split tube to guide the first cut is desirable.

The proper degree of dryness to bring out maximum contrast in color varies considerably with different soils and can be found only by trial. Therefore, a technician should check the sample as it dries in order

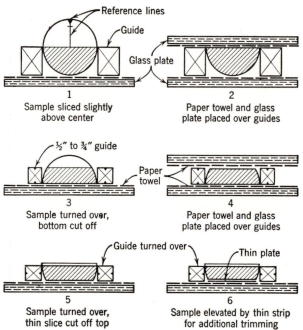

FIGURE I–4. Procedure in slicing small samples. (From "Subsurface Exploration and Sampling of Soils for Civil Engineering Purposes," a project of the Engineering Foundation. See reference I–7.)

to determine the proper stage of dryness. If the sample is dried past the peak point, it can often be returned to the peak point by slight moistening. The rate of drying should be slow enough to prevent the soil from cracking; it can be retarded by the use of moist towels, wax paper, storage in a humid room, etc.

Because during the original slicing the strata are often smeared, the surface of the drying clay should be trimmed with a very sharp knife just before photographing. Care should be taken to keep the trimmed surface smooth since grooves, etc., show up in photographs.

Several partially dried specimens are shown in Fig. I–5. Figure I–5a shows the distortion of the soil around a stone; Fig. I–5b is a tested unconfined compression [10] specimen; Fig. I–5c is a quarter section of a triaxial compression [10] specimen showing the failure

[10] This test is presented in a later chapter.

plane; and Fig. I–5d is a section showing a mixing of strata, probably caused by a disturbance during deposition of the soil or a landslide. The small holes in the lower left corner of Fig. I–5d were occupied by sand grains which fell out upon drying.

Water Content

Among the most frequently determined soil characteristics is water content. The water content, w, of a soil mass is defined as the ratio (usually expressed as a percentage) of the weight of water to the weight of dry soil grains in the mass, or the equation [11]

$$w = \frac{W_1 - W_2}{W_2 - W_c} \qquad (I-1)$$

in which W_1 = weight of container plus moist soil,
W_2 = weight of container plus oven-dry soil,
W_c = weight of container.

Although the water content is one of the easiest properties of a soil to obtain, it is also one of the most useful. For example, much research has shown that water content is a good indication of the shear strength of a saturated clay.

In selecting a sample for a water content determination, we must exercise care to obtain a representative specimen. In sedimentary clays there may be a major difference in water content between adjacent strata (for example, see Fig. I–6); also, many soil samples are likely to have dried at the surface. Therefore, if a water content is to be determined which is representative of a test specimen, soil for the determination should be obtained from more than one stratum of material which has not undergone surface drying.

The amount of soil taken for a water content specimen depends on the type of soil and the quantity available. In general, the larger the specimen, the more accurate is the determination because the weights involved are larger. Often very small specimens of clay are taken to obtain some indication of water content variation. For example, Fig. I–6 is a plot of water content against depth for a foot-long vertical section of Boston blue clay (I–2), in which the water contents were determined every ¼ in. in depth. Figure I–6 shows a range of water content from 25% to 57% within a depth of 2 in.

A water content specimen should be weighed as soon as possible after it has been secured to minimize the effects of drying. Watch glasses properly clipped

[11] A derivation of every formula used in this book is given in Appendix B unless it is derived in the main body of the text where it is used, or otherwise noted.

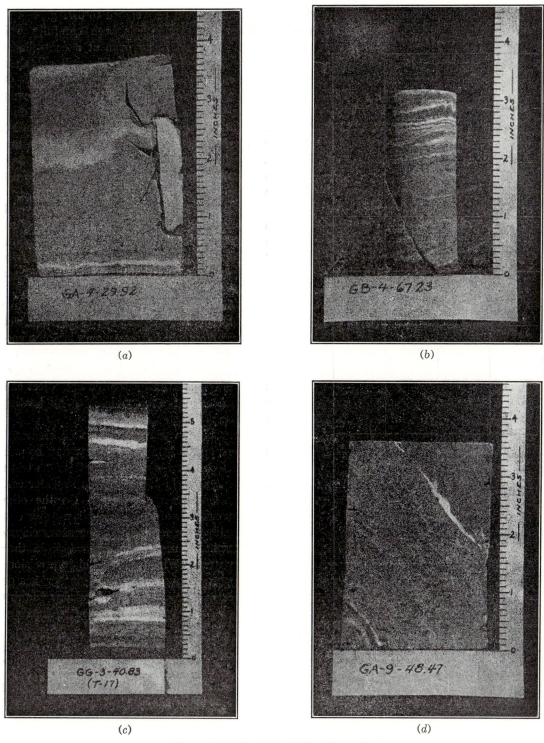

FIGURE I–5. Partially dried soil samples.

together (i.e., no soil on the ground edges) will prevent escape of water. Metal cans, although they are more convenient to use than watch glasses, have the disadvantage of not completely preventing soil drying and of possibly changing weight because of corro-

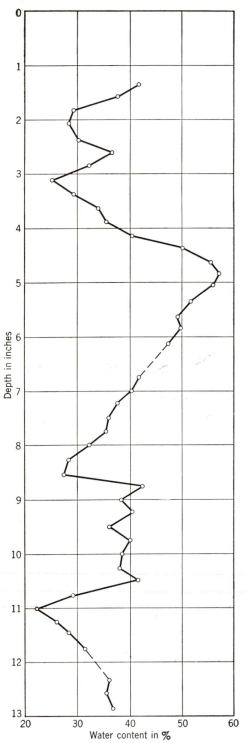

Depth in inches

Water content in %

FIGURE I–6. Water content versus depth of Boston blue clay. (Data for figure from reference I–2.)

sion.[12] For most testing, however, cans are acceptable. The balance used for weighing should be selected in keeping with the size of the soil specimen, e.g., a 10-g specimen should be weighed to 0.01 g.

After the sample has been carefully weighed, it should be dried in an oven at a controlled constant temperature of 105° C.[13] The soil container should be opened when it is placed in the oven. Often the sample is dried overnight in the oven, even though the actual drying time necessary depends on the type, amount, and shape of soil specimen used. For example, a few grams of sand can be dried in an hour or less, whereas the same weight of a very fine-grained clay may require many hours to come to a constant weight.

After a container of soil has been removed from the oven, it should be cooled, then weighed. A nonplastic soil may be cooled normally at room humidity. If a plastic soil is weighed within an hour or two after it has been removed from the oven, cooling at room humidity is satisfactory for routine testing. On the other hand, a plastic soil should be kept in a desiccator if the period [14] between drying and weighing is longer. Weighing hot containers is not a good policy for two reasons. First, one is more likely to drop or spill the contents of a hot container than a container at room temperature; second, a hot object can disturb the accuracy of a beam balance by heating it unsymmetrically.

Because of variations in water content between adjacent strata in a sedimentary clay as pointed out before (see Fig. I–6), attempts to refine test methods in order to obtain high precision in water content determinations are inconsistent. On the other hand, high precision is often wanted when water content determinations are made on homogeneous clays or when the weight of dry soil is measured in the specific gravity test. When seeking high precision, one is faced with the question, "Just how 'dry' is a dry soil?" Most engineering specifications define the dry weight of a soil as that weight obtained by heating the soil

---

[12] The change in the weight of a can because of corrosion during a single water content determination is negligible. A convenient procedure is to weigh all containers and post a list of their weights near the balance. The weight of any one may then be read from the list. If such a procedure is followed, the effect of corrosion on the weight of a can becomes an important consideration.

[13] Temperatures of 105° C and 110° C are both commonly used as drying temperatures. The difference of 5° C is of little practical importance.

[14] An example of a high rate of water pick-up at a relative humidity of 50% to 60% is furnished by a sample of Mexico City clay which adsorbed 0.6% moisture in 2 hours and 2.2% in 24 hours (I–8).

at 105° C (or 110° C) until the weight reaches a constant value. These specifications incorrectly imply that all the water is driven from the voids or pores of the soil while the soil particles remain unaltered in composition. Actually the effective soil particle consists of the grain plus an adsorbed layer of water,

distributing ovens controlled to a temperature of 110° C revealed that "temperatures in different parts of the oven chamber varied, as in a typical case, from 99.6 to 146.7° C, a variation of 47.1° C" (I–5). Figure I–8 is a plot of depth against temperature within three nonheat-distributing ovens that were used

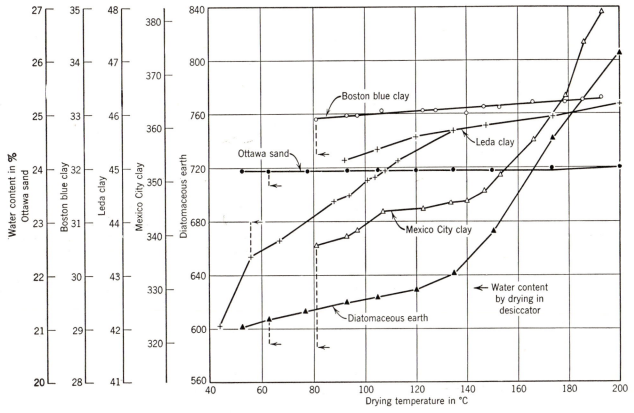

FIGURE I–7. Drying curves. (From reference I–8.)

which can be completely driven off only by heat much greater than 105° C.

In Fig. I–7 (I–8) are presented drying curves for five soils which are extremely different in their characteristics. These curves show that there is nothing special about 105° C which makes its selection as a drying temperature scientific. Also in Fig. I–7 are shown the water contents obtained by drying in a standard desiccator over calcium chloride. As would be expected, the drying curves of soils which are colloidal in behavior show the most variation.

Figure I–7 shows that, in order to obtain consistent dry soil weights by oven drying, a technician must carefully control the temperature of the oven. Few people realize the magnitude of temperature variation which exists in some of the commonly used "constant-temperature" ovens. Central Scientific Company reports that carefully conducted tests on nonheat-

in a soil mechanics laboratory. These data were obtained from observations on thermometers placed on shelves of the ovens "controlled" at 105° C.[15]

As indicated above, the dry weight of a soil is dependent on the method of drying; therefore, such soil properties as void ratio, porosity, degree of saturation, specific gravity, and grain size distribution described in the following pages are influenced by the method of drying since they depend on dry soil weight. The drying method which drives off more water tends to give the higher value of void ratio and porosity. The degree of saturation is only slightly affected by the drying method. The effect of drying method on specific gravity is discussed on page 18 and on grain size distribution on page 35.

[15] Note that the ordinary oven controlled at 105° C may reach temperatures high enough to decompose some types of organic matter which occur in soils.

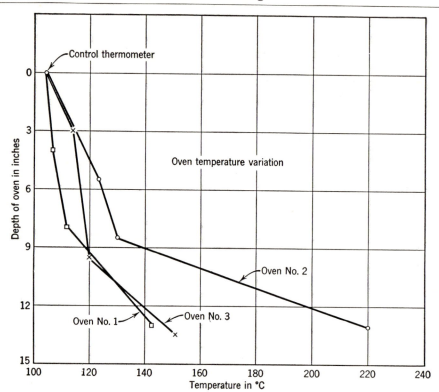

FIGURE I–8. (From reference I–8.)

Void Ratio

The void ratio, e, is defined as the ratio of void volume [16] to solid volume in a soil mass, and may be found from

$$e = \frac{G\gamma_w V}{W_s} - 1 = \frac{V_v}{V_s} \qquad (I\text{–}2)$$

in which G = specific gravity of soil solids (see Chapter II),

γ_w = the unit weight of water (Table A–2, p. 147),

V = volume of soil mass,

W_s = dry weight of soil grains.

The void ratio is a measure of the denseness of a soil and is, therefore, one of the most revealing characteristics of a soil. Several soil properties, such as permeability and strength, are related [17] to void ratio.

To determine a void ratio the variables in Eq. I–2 must be obtained. The test procedure for getting the specific gravity is given in Chapter II; the unit weight of water at any particular temperature is easily acquired from Table A–2, page 147; and the method of

[16] The void volume of a soil mass is any volume not filled with particles. Therefore, the void volume plus the solid volume, or volume of particles, is equal to the total volume.

[17] These relationships are discussed in later chapters.

securing the weight of dry soil was presented in the preceding section of this chapter.

The volumes of specimens for shear tests are usually computed from measured dimensions. Because these specimens only approach special shapes such as cylinders and because precise measurements of length are hard to get, a high degree of precision in their volume determinations is not common. Better volume measurements are obtained when soil is placed in a container of known dimensions, as is done in the compaction, permeability, consolidation, and capillary tests. Likewise, the volume of a small, smooth sample of cohesive soil can be measured with a fair degree of precision by means of the mercury displacement method described on page 25 for the shrinkage limit pat.

Even though knowledge of the void ratio of a soil is more important than knowledge of the water content, far more water content determinations are made on shear specimens of cohesive soils. The reasons that fewer void ratio measurements are made are that they are more difficult to do and that void ratio is proportional to water content for a given saturated soil (see Eq. I–5).

Porosity

The porosity, n, defined as the ratio of void volume to total volume in a soil mass, may be found from

$$n = 1 - \frac{W_s}{G\gamma_w V} \qquad (I\text{-}3)$$

Like void ratio, porosity is a measure of the denseness of a soil. The computation of porosity involves the same measurements as the computation of void ratio. The two are uniquely related by

$$n = \frac{e}{1 + e} \qquad (I\text{-}3a)$$

As a soil compresses or swells, the void volume and the total volume change while the volume of solids remains constant. Thus the porosity, V_v/V, is a fraction of which both the numerator and denominator are variables, whereas void ratio, V_v/V_s, has a constant denominator and variable numerator. Because of this fact, void ratio is a more convenient measure of volumetric strain than the porosity is. Porosity, on the other hand, is more convenient than void ratio in seepage problems.

Degree of Saturation

The degree of saturation, S, is the ratio of the volume of water in the soil voids to the total volume of voids. Since the degree of saturation is usually expressed as a percentage, it is often called percentage of saturation. It can be found from

$$S = \frac{W_w}{\gamma_w V_v} \qquad (I\text{-}4)$$

in which W_w = the weight of water,
V_v = the volume of voids = $V - V_s$.

Knowledge of the degree of saturation is extremely useful because this characteristic influences such fundamental soil properties as permeability, shear strength, and compressibility.[18] With the exception of V_s, which is computed from $V_s = (W_s/G\gamma_w)$, the values needed to solve Eq. I–4 have been discussed in the preceding pages. Unfortunately, the degree of saturation cannot always be obtained with precision since it, like the void ratio, depends on volume measurements.

Below is an equation which relates S, G, e, and w; it is often useful for checking arithmetic.

$$Se = Gw \qquad (I\text{-}5)$$

Report of Laboratory Test

The nature of a good report on a soil test depends on the intended use of the report. On the one hand, a report made to a structural engineer to furnish him

[18] These relationships are discussed in later chapters.

with soil data needed for the design of a structure should present the test results clearly and concisely. The test procedure employed should be noted either by a brief description or by a reference to some published procedure. Often the results for a report to a structural engineer can be given in the form of a plot and/or a summary table. Methods of presenting results are discussed for each of the tests covered in the following pages of this book.

On the other hand, a student report, the purpose of which is to teach the fundamentals of the soil test, might include:

1. *Procedure.* A reference to the procedure followed. Any steps of the procedure referred to which were not actually followed should be noted.

2. *Data.* All data recorded in the laboratory. The use of carefully organized data sheets, such as those employed in the numerical examples given in the following pages, is recommended. A student should strive to learn to take neat and complete laboratory data since copying data is a waste of time and invites errors.

3. *Calculations.* A sample of all types of calculations with a reference to every formula employed.

4. *Results.* A presentation of the test results, usually in the form of a summary table and/or curves.

5. *Discussion.* In the discussion, the student may:
   (a) Discuss any difficulties experienced or inaccuracies introduced in performing the test.
   (b) Suggest any improvements in the procedure employed.
   (c) Account for any unusual or unexpected results.
   (d) Make a statement concerning the probable accuracy of the results.
   (e) Compare the results obtained with those for other soils.

The preparation of a report covering the five items suggested above requires a sound understanding of the soil test. Unfortunately, approximately five hours are needed to prepare such a report for most of the soil tests. The laboratory instructor, therefore, may have to settle for a less complete report because of the limitation of time.

REFERENCES

1. Abdun-Nur, "Classification of Soils as Used by the Bureau of Reclamation," presented before the American Society for Testing Materials, June, 1950.

2. Albin, Pedro, Jr., "Laboratory Investigation of the Variable Nature of the Blue Clay Layer below Boston," Master of Science Thesis, Massachusetts Institute of Technology, May, 1947.

3. American Association of State Highway Officials, *Standard Specifications for Highway Materials,* 1950.

4. Casagrande, A., "Classification and Identification of Soils," *Transactions of the American Society of Civil Engineers,* Vol. 113, p. 901, 1948.

5. *Catalog J136,* Central Scientific Company, Chicago, Illinois,

6. Housel, William S., "Shearing Resistance of Soil," *University of Michigan Reprint Series* No. 13, May, 1940.

7. Hvorslev, J. M., "Subsurface Exploration and Sampling of Soils for Civil Engineering Purposes," Waterways Experiment Station, Vicksburg, Miss., November, 1948.

8. Lambe, T. W., "How Dry Is a 'Dry' Soil?", *Proceedings of the Highway Research Board,* 1949.

9. Mohr, H. A., "Exploration of Soil Conditions and Sampling Operations," *Harvard University Soil Mechanics Series* No. 9, February, 1940.

10. Osterberg, J. O., and K. H. Tseng, "The Effectiveness of Various Waxes for Sealing Soil Samples," unpublished.

11. Rutledge, P. C., "Description and Identification of Soil Types," *Proceedings of the Purdue Conference on Soil Mechanics and Its Applications,* July, 1940.

12. Rutledge, P. C., "Relation of Undisturbed Sampling to Laboratory Testing," *Transactions of the American Society of Civil Engineers,* 1944.

13. Waterways Experiment Station, "Preservation of Sliced Soil Samples," *Technical Memorandum* No. 3–306, Vicksburg, Miss., December, 1949.

14. Wilson, S. D., and R. W. Brandley, "Improved Technique Saves Times in Obtaining Undisturbed Chunk Samples of Clay," *Civil Engineering,* April, 1948.

CHAPTER

II

Specific Gravity Test

Introduction

The specific gravity of a soil is the ratio of the weight in air of a given volume of soil particles to the weight in air of an equal volume of distilled water at a temperature of 4° C. The specific gravity of a soil is often used in relating a weight of soil to its volume. Thus, knowing the void ratio, the degree of saturation, and the specific gravity, we can compute the unit weight of a moist soil. Unit weights are needed in nearly all pressure, settlement, and stability problems in soil engineering. The specific gravity is also used in the computations of most of the laboratory tests.

Although specific gravity is employed in the identification of minerals, it is of limited value for identification or classification of soils because the specific gravities of most soils fall within a narrow range.

Apparatus and Supplies

Special
  1. Pycnometer (volumetric bottle)
General
  1. Distilled water
  2. Vacuum source (optional)
  3. Heat source (such as a burner or hot plate)
  4. Balance (0.01 g sensitivity) [1]
  5. Drying oven
  6. Desiccator
  7. Thermometer (graduated to 0.1° C)
  8. Evaporating dishes
  9. Medicine dropper or pipette

Figure II–1 shows two pycnometers.[2] The one on the right, which is more commonly used in soil laboratories than the other, has a calibrated capacity of

[1] A more sensitive balance may be required if a small pycnometer is used.

[2] Actually the large one is a volumetric bottle and not a pycnometer.

500 ml at 20° C within a tolerance of 0.30 ml. Both are made of pyrex glass. It is easier to pour soil into the large bottle; however, the small one has a thermometer in the neck, the tip of which is regulated to the proper depth for recording the temperature.

The detailed test procedure and the numerical example which follow employ the large and more common volumetric bottle.

Recommended Procedure [3]

In the computation of the specific gravity of a soil from laboratory data, the weight of the pycnometer filled with distilled water at the test temperature will be needed. This value is usually taken from a plot of temperature versus weight of bottle plus water. The plot, or calibration curve, can be determined either by experimental or by theoretical means:

I. *Bottle Calibration*

  A. *Experimental Procedure*

This procedure consists of obtaining at least three sets of concurrent temperature and weight measurements about 4° C apart and within the temperature range of 20° to 30° C. Each set, representing the coordinates for a point on the calibration curve, is obtained as follows:

  1. To a clean pycnometer add deaired,[4] distilled water at room temperature until the bottom of the meniscus is at the calibration mark.

[3] A student doing this test for the first time should be able to calibrate one bottle and make a specific gravity determination on one soil in 2 to 3 hours. A student group can split into two parts, one doing one type of calibration and the specific gravity determination on one type of soil and the other doing the other calibration and determination. At the end of the test the two parts should combine their data for the preparation of their reports.

[4] Truly deaired water is water in which there is no air dis-

2. Carefully dry the outside of the bottle and the inside of the neck above the water surface.

3. Weigh the bottle plus water to 0.01 g.

4. Measure the water temperature to 0.1° C. Hold

The recorded temperature is taken with the thermometer inserted to the depth at which the thermometer is designed to read. (This depth is usually marked on the thermometer.)

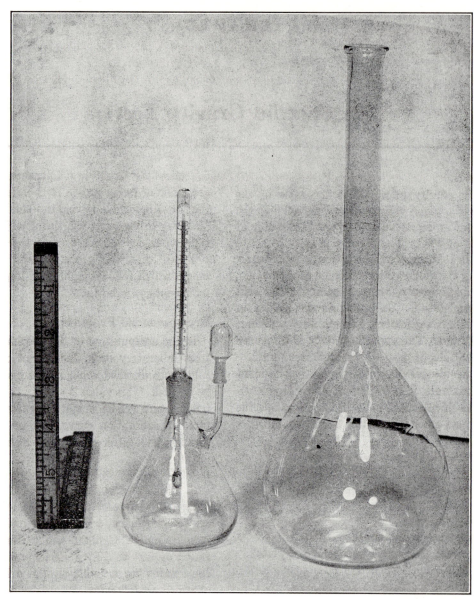

FIGURE II–1. Pycnometers.

the tip of the thermometer at different elevations within the water to see if the temperature is uniform.

solved; however deaired is often loosely used for water which has less than its capacity of dissolved air. If deaired water is not used in step 1, air may come out of solution and appear as objectionable bubbles when the water is warmed in step 6. The effect of dissolved air on the density of water is negligible as evidenced by the following: "Between the temperatures 5° C and 8° C the density of water saturated with air was found to be 0.0000030 g/ml less than the density of air-free water" (II–3, page 26).

5. If the temperature is nonuniform, place the thumb over the open end of the bottle and turn it upside down and back to mix the water thoroughly for a temperature observation.

6. Heat the bottle of water slightly by placing it in a warm water bath and repeat steps 2–5, each time removing enough water to bring the meniscus down to the calibration mark. Repeat this procedure until enough points are obtained to plot the calibration curve.

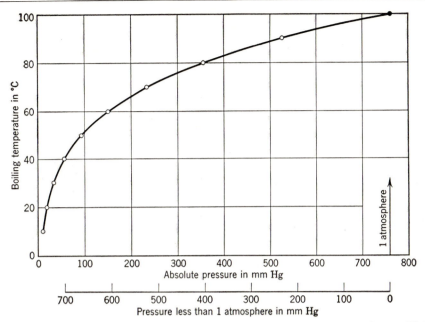

FIGURE II-2. Boiling temperature of water versus pressure. (Data from reference II-3.)

B. Theoretical Procedure

Points for the calibration curve can be obtained by successively substituting different temperatures in the following equation:

$$W_2 = W_B + V_B(1 + \Delta T \cdot \epsilon)(\gamma_T - \gamma_a) \quad \text{(II-1)}$$

in which W_2 = weight of bottle plus water,

W_B = weight of clean, dry bottle,

V_B = calibrated volume of bottle at T_c,

$\Delta T = T - T_c$,

T = temperature in degrees centigrade at which W_2 is desired,

T_c = calibration temperature of bottle (usually 20° C),

ϵ = thermal coefficient of cubical expansion for pyrex glass = 0.100×10^{-4} per °C,

γ_T = unit weight of water at T (see Table A-2, p. 147),

γ_a = unit weight of air at T and atmospheric pressure (an average value of γ_a accurate enough for use in this test is 0.0012 g per cc).

The only bit of laboratory data needed is the weight of the pycnometer, which must be absolutely clean and dry.[5] A good practice is to obtain one check point for the calibration curve by the previously described experimental method. (See Numerical Example.)

II. Specific Gravity Determination

A. Cohesionless Soil

1. Put approximately 150 g of oven-dry soil, weighed to 0.01 g, into a calibrated pycnometer which is already half full of deaired, distilled water. Be sure no soil grains are lost when they are put into the pycnometer.

2. Remove all of the air [6] which is entrapped in the soil by 10 minutes of boiling;[7] accompany the boiling with continuous agitation. The application of a partial vacuum to the suspension of soil in water to lower the boiling temperature is desirable, since the lower the temperature at which the suspension is boiled, the less the cooling which will have to be done later.[8] Figure II-2, a plot of boiling temperature of pure water [9] against applied pressure, indicates the effect of reduced pressure on boiling temperature.

3. Cool the bottle and suspension to some temperature within the range of the calibration curve for the bottle.

---

[5] To clean and dry a pycnometer, first wash it thoroughly and allow it to drain. Rinse it with alcohol to remove the water and then drain the alcohol. Next rinse with ether to remove the alcohol and then drain the ether. Turn bottle upside down for a few minutes to permit the ether vapor to come out.

[6] The presence of entrapped air can be detected by the movement of the surface of the suspension upon the application and release of the vacuum.

[7] The boiling should not be too vigorous since soil may be carried out the neck of the bottle.

[8] The heating of a pycnometer to a high temperature is unwise because of the possibility of changing the shape of the bottle.

[9] The presence of foreign matter, such as soluble salts from the soil, raises the boiling point slightly.

4. Add water to bring the bottom of the meniscus to the calibration mark.

5. Dry the outside of the bottle and the inside of the neck above the meniscus.

6. Weigh the bottle with water and soil in it to 0.01 g.[10]

7. After checking to be sure that the contents of the bottle are at a uniform temperature, record[11] the temperature. (See step 4 of the experimental calibration.)

B. Cohesive Soil

1. Work a sample of the soil to be tested into a smooth paste by mixing it with distilled water. The sample used should contain approximately 50 g in dry weight.

2. Pour the paste into a calibrated pycnometer.

3. Remove the entrapped air;[12] cool, and obtain the weight and temperature as was done in steps 2–7 of the procedure for the cohesionless soil.[13]

4. Pour the entire mixture of soil and water into a large evaporating dish of known weight; rinse the pycnometer carefully to insure the collection of all the soil.

5. Dry the soil in the oven, cool, and weigh. The dry weight of soil grains can be obtained by subtracting the weight of the empty dish from the weight of the dish with soil in it.

Discussion of Procedure

Even though the preceding procedures are basically simple, a technician must exercise the utmost care in obtaining temperature and weight measurements in order to secure reasonably accurate results. Such care is necessary because Eq. II–2 contains a difference of weights which is small in comparison with the weights themselves. In cases like these, a small error in either weight may be very significant. A part of the error inherent in the balance can be eliminated

[10] Better accuracy is obtained if the same balance and weights are used here as were used in the calibration, since a difference of weights, which tends to cancel some of the error of the scale, is employed in Eq. II–2. The hundredths will have to be estimated on a balance with capacity enough for the 500-ml bottle.

[11] A recommended procedure is to obtain several sets of temperature and weight observations as the sample cools (see Numerical Example). The most accurate set of readings is usually obtained at room temperature, since a more uniform temperature is likely to exist.

[12] See footnote 6.

[13] Often the soil-water suspension is opaque, which makes it difficult to bring the bottom of the meniscus exactly to the calibration mark. The use of a strong light behind the bottle neck helps in seeing the meniscus bottom.

by using the same instrument for every weighing, as suggested in footnote 10.

In a soils laboratory each pycnometer should be carefully calibrated and its calibration curve (see page 20) kept available so that calibration is not necessary each time a specific gravity determination is made. A curve may be checked from time to time since such things as scratching or dirt accumulation may change the weight of the bottle.

The theoretical method of bottle calibration does not lend itself to a high degree of precision because of the large tolerance in calibrated volume of the bottle (the volume of a bottle can be determined in the laboratory) and the difficulty in obtaining precisely the dry weight and the cubical coefficient of expansion of the bottle. The experimental method, therefore, is recommended in normal soil testing.

Two sources of important experimental errors are nonuniform temperatures and incomplete removal of air entrapped in the soil. Nonuniform temperature difficulties can easily be prevented by allowing a warmed bottle to stand overnight to adjust itself to room temperature. The boiling procedure specified in step 2, page 17, is normally sufficient to remove air within the soil.

As pointed out on page 11, the effective soil particle contains a film of adsorbed water; therefore, the specific gravity obtained is dependent on the method of drying employed. As an indication of the possible variation that can be obtained, the specific gravities of five soils by different drying methods (II–4) are listed in Table II–1. This table shows that careful

TABLE II–1

| Soil | Specific Gravity | | | | |
Drying Method	Desiccator	90° C	105° C	140° C	190° C
Ottawa sand	2.67	2.67	2.67	2.67	2.67
Diatomaceous earth	1.91	1.99	2.00	2.08	2.56
Boston blue clay	2.76	2.78	2.78	2.78	2.79
Mexico City clay	2.22	2.33	2.35	2.37	2.62
Leda clay	2.74	2.75	2.77	2.80	2.82

control of the drying method may be necessary if consistent values of specific gravity are to be obtained.

Calculations

The specific gravity[14] of the soil, G_s, can be obtained from

$$G_s = \frac{W_s G_T}{W_s - W_1 + W_2} \qquad (II-2)$$

[14] The symbol G without subscript is commonly used for the specific gravity of soil solids where there is no chance of its being mistaken for some other specific gravity

in which G_T = specific gravity of distilled water at temperature T (see Table A–2, p. 147),

W_s = dry weight of soil,

W_1 = weight of pycnometer, soil, and water,

W_2 = weight of pycnometer plus water (from calibration curve).

Results

Method of Presentation. A pycnometer calibration is usually presented as a plot of temperature versus weight of bottle plus water (see page 20). As for the specific gravity obtained, a statement of the value is normally a sufficient presentation of the test results.

Typical Values. The specific gravity of most soils lies within the range of 2.65 to 2.85. Soils with measurable organic content or soils with porous particles, such as diatomaceous earth, may have specific gravity values below 2.0.[15] On the other hand, soils containing heavy substances, such as iron, may have specific gravity values above 3.0. An indication of specific gravities can be obtained from the middle column in Table II–1.

---

[15] It may be more correct to grind up porous particles for a specific gravity test and thus obtain a much higher value for the specific gravity. In such a case, the actual degree of saturation of the soil in situ should be considerably lower than normally determined.

The specific gravity of a natural soil often depends on how representative the test specimen is, because most soils are far from homogeneous. For example, Boston blue clay often contains lenses of silt which may cause the specific gravity to vary from 2.70 to 2.80, depending on the percentage of silt in the test specimen.

Numerical Example [16]

The example on pages 20 and 21 presents a pycnometer calibration obtained by the theoretical procedure. One experimental point is also plotted on the curve. The soil of the test was a glacial till used for the core of an earth dam.

REFERENCES

1. *Catalog,* Eimer and Amend, 625 Greenwich Street, New York, New York.
2. *Catalog,* Fisher Scientific Company, 717 Forbes Street, Pittsburgh, Pa.
3. *International Critical Tables,* McGraw-Hill Book Co., Vol. III, 1928.
4. Lambe, T. W., "How Dry Is a 'Dry' Soil?", *Proceedings of the Highway Research Board,* 1949.

---

[16] The Numerical Example is intended as a guide to the student.

SOIL MECHANICS LABORATORY

PYCNOMETER CALIBRATION

PYCNOMETER NO. _____8_____   TESTED BY _WCS_____

                                         DATE _____June 24, 1950_____

A. EXPERIMENTAL PROCEDURE

DETERMINATION NO.	1	2	3	4
WT. BOTTLE + WATER, W_2, IN g	674.55			
TEMPERATURE, T, IN °C	22.3			

B. THEORETICAL PROCEDURE

WT. BOTTLE, W_B, IN g	176.37	CUBICAL EXPANSION FOR GLASS IN 1/°C, ϵ,	$.100 \times 10^{-4}$
TEMPERATURE OF CALIBRATION, T_c, IN °C	20.0	UNIT WEIGHT OF AIR γ_a, IN g/cc	0.0012
VOL. BOTTLE AT T_c, V_B, IN cc	500.00	$W_2 = W_B + V_B(1 + \Delta T \cdot \epsilon)(\gamma_T - \gamma_a)$	

DETERMINATION NO.	1	2	3	4
TEMPERATURE, T, IN °C	20.0	25.0	30.0	35.0
UNIT WEIGHT OF WATER AT T, γ_T, IN g/cc	0.9982	0.9971	0.9957	0.9941
WT. BOTTLE, + WATER AT T, W_2, IN g	674.87	674.35	673.67	672.89

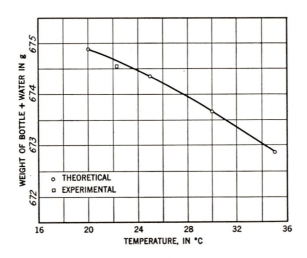

SOIL MECHANICS LABORATORY

SPECIFIC GRAVITY TEST

SOIL SAMPLE _Sandy-Clayey Silt: grayish brown at_
natural water content; well graded subrounded
particles; mostly quartz & feldspar, some mica;
some fine roots; glacial till (boulder clay)

TEST NO. _SG-6_

DATE _June 25, 1950_

LOCATION _Union Falls, Maine_

TESTED BY _WCS_

BORING NO. _____ SAMPLE DEPTH _2.0 ft._

SAMPLE NO. _F9_

DETERMINATION NO.	1	2	3	4
BOTTLE NO.	8	8	8	
WT. BOTTLE + WATER + SOIL, W_1, IN g	706.11	706.80	707.07	
TEMPERATURE, T, IN °C	30.0	25.5	23.0	
WT. BOTTLE + WATER, W_2, IN g	673.67	674.28	674.57	
EVAPORATING DISH NO.	A-15	A-15	A-15	
WT. DISH + DRY SOIL IN g	491.12	491.12	491.12	
WT. DISH IN g	438.92	438.92	438.92	
WT. SOIL, W_s, IN g	52.20	52.20	52.20	
SPECIFIC GRAVITY OF WATER AT T, G_T	0.996	0.997	0.998	
SPECIFIC GRAVITY OF SOIL, G_s	2.63	2.64	2.64	

REMARKS

$$G_s = \frac{G_T W_s}{W_s - W_1 + W_2} \; ; \qquad G_s \; \underline{2.64}$$

CHAPTER

III

Atterberg Limits and Indices

___

Introduction

A fine-grained soil can exist in any of several states; which state depends on the amount of water in the soil system. When water is added to a dry soil, each particle is covered with a film of adsorbed water. If the addition of water is continued, the thickness of the water film on a particle increases. Increasing the thickness of the water films permits the particles to slide past one another more easily. The behavior of the soil, therefore, is related to the amount of water in the system. Approximately forty years ago, A. Atterberg (III–5) defined the boundaries of four states in terms of "limits" as follows: (a) "liquid limit," the boundary between the liquid and plastic states; (b) "plastic limit," the boundary between the plastic and semi-solid states; and (c) "shrinkage limit," the boundary between the semi-solid and solid states. These limits have since been more definitely defined by A. Casagrande (III–6) as the water contents which exist under the conditions described in the following pages.

The liquid limit is the water content at which the soil has such a small shear strength [1] that it flows to close a groove of standard width when jarred in a specified manner. The plastic limit is the water content at which the soil begins to crumble when rolled into threads of specified size. The shrinkage limit is the water content that is just sufficient to fill the pores when the soil is at the minimum volume it will attain by drying. The amount of water which must be added to change a soil from its plastic limit to its liquid limit is an indication of the plasticity of the soil. The plasticity is measured by the "plasticity index," which is equal to the liquid limit minus the plastic limit.

Although the liquid and plastic limits are necessarily determined on soils which have had their natural structure completely destroyed by kneading or "remolding," the shrinkage limit can be obtained on soils in either their undisturbed or their remolded states. The difference between the undisturbed and remolded shrinkage limits may be an indication of the amount of natural "structure" [2] a soil possesses. Also the condition of an in situ soil is often partially revealed by its "water-plasticity ratio," [3] which is the ratio of the difference between the natural water content and the plastic limit to the plasticity index. A high water plasticity ratio, which means that the natural water content is high relative to the liquid limit, indicates a very low remolded strength. For example, if the ratio is greater than 100%, [4] the soil exists at a water content greater than the liquid limit, and its remolded strength is thus less than that very small amount which it would possess at the liquid limit.

The chemical and mineral composition, size, and shape of the soil particles influence the adsorbed water films on the particles. Because such soil properties as compressibility, permeability, and strength, [5] as well as the limits, are dependent on the water films,

[1] This strength is a definite value; therefore, all soils have the same strength at their respective liquid limits.

[2] The word "structure" has a special meaning in soil mechanics. A soil is said to have structure if an undisturbed sample has strength which is lost with kneading. Structure is further discussed on page 111.

[3] This ratio is sometimes called the "liquidity index" (III–14).

[4] A soil which has a water-plasticity ratio greater than a 100% is termed "quick-clay" by Norwegian engineers (III–11).

[5] These properties are discussed in later chapters.

approximate relationships exist between these properties and the limits. Some general relationships between the limits and engineering properties as given by A. Casagrande (III–7) are in Table III–1.

TABLE III–1

Characteristic	Comparing Soils at Equal Liquid Limit with Plasticity Index Increasing	Comparing Soils at Equal Plasticity Index with Liquid Limit Increasing
Compressibility	About the same	Increases
Permeability	Decreases	Increases
Rate of volume change	Decreases	....
Toughness near plastic limit	Increases	Decreases
Dry strength	Increases	Decreases

For a given soil, we can often set up definite semi-empirical relationships between a property and the limits or indices. From such expressions, we can then make predictions of the properties of another sample of the same soil by knowing the limits. Such a procedure is often very helpful because the limits are usually more easily determined than the compressibility, permeability, or strength.

On construction jobs in which detailed studies of the underlying soil conditions are made, plots of limits, along with other test results, against depth can be made. For example, Fig. XII–8 presents such a plot from a boring in which a fixed-piston sampler was employed to obtain undisturbed soil samples. From this plot, we can see that the samples with the lowest strength had the highest water-plasticity ratios. Because of this fact, water-plasticity ratios were employed to help locate pockets of weak clay at this site. The ratios could be determined from spoon samples which are more easily obtained than fixed-piston samples.

The limits furnish an excellent basis for the classification [6] and identification [7] of fine-grained soils. They are also often used directly in specifications for controlling soil for use in fill, and in semi-empirical methods of design. For example, the design of flexible pavements by the Civil Aeronautics Administration method is partly based on the limits (III–10).

From the preceding discussion we can see that, even though the limits do not furnish numbers which can be substituted directly into scientifically derived formulas, they are extremely useful to the soil engineer.

Apparatus and Supplies

Special

1. Liquid limit device and grooving tool
2. Shrinkage limit set, consisting of:
   (a) Petri dish
   (b) Glass plate (with prongs)
   (c) Mercury supply
   (d) Large evaporating dish
   (e) Medicine dropper
3. Large glass plate for plastic limit

General

1. Distilled water
2. Balances (0.01 g sensitivity and 0.1 g sensitivity)
3. Drying oven
4. Desiccator
5. Watch glasses or drying cans
6. Evaporating dishes
7. Spatula

The liquid limit device is shown in Figs. III–1, III–2, and III–3; the plastic limit plate in Figs. III–4 and III–5; and a diagrammatic sketch of the shrinkage limit set in Fig. III–6. The liquid limit device is operated by turning the crank which raises the cup and lets it fall. The early devices were made with a hard rubber base; however, during World War II, wood was substituted because of the scarcity of rubber. As we might expect from the difference in the restitutional properties of rubber and wood, liquid limit determinations made on the two types of devices do not agree. Research [8] has shown that Micarta [9] can be successfully used for bases. On pages 152 and 153, Appendix A, are detailed drawings of the liquid limit device prepared at the Soil Mechanics Laboratory at Harvard University. These detailed drawings show rubber feet on the base. These feet cushion the base and help prevent its rocking on the table. Prior to the installation of feet, something such as a newspaper was used between the base and table.

The distance which the cup drops, 0.394 in., must be properly set and checked at frequent intervals. It is the height of fall of the point on the cup which strikes the base; this point is not the lowest point on the cup when the cup is raised. (See page 152, Appendix A.)

[6] The limits are essential to the Airfield Classification System (III–7).

[7] It has been suggested that the limits be the sole criterion of the boundary between clay and silt. Soil with particles finer than 0.074 mm would be called a silt if its liquid limit was 28% or less and its plasticity index 6% or less; it would be called clay if its liquid limit was over 28% and plasticity index over 6%.

[8] At the Harvard Soil Mechanics Laboratory.

[9] Trade name for a plastic made by the Westinghouse Company.

Recommended Procedure [10]

Since the water-plasticity ratio requires the natural water content, we should first obtain the natural water content of the soil sample by the procedure given in Chapter I. The limits should be determined on that portion of the soil finer than a No. 40 sieve.[11] If the test specimen contains clay, it should never have been drier than approximately its plastic limit.

I. Liquid Limit Determination [12]

1. Take about 100 g of moist soil and mix it thoroughly with distilled water to form a uniform paste.

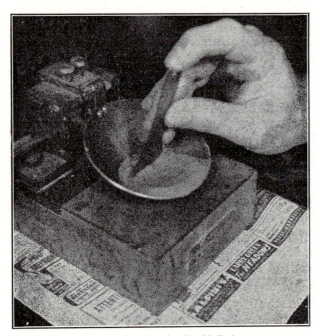

FIGURE III–1. Cutting the liquid limit groove.

2. Place a portion of the paste in the cup of the liquid limit device, smooth the surface off to a maximum depth of ½ in., and draw the grooving tool [13]

[10] A student performing this test for the first time should be able to determine the limits and indices presented in this chapter for one soil in 2 to 3 hours. He should be able to do the computations in about an hour.

[11] To run the limits on soil finer than a No. 140 or No. 200 sieve appears more logical. The No. 40 sieve, however, is commonly accepted as the upper limit of grain sizes used for the limits.

[12] A technician can obtain more consistent values if he performs this test in a humid atmosphere, such as that in a humid room. A humid atmosphere minimizes surface drying of the soil while it is being tested.

[13] The grooving tool shown in Fig. III–1 and in Appendix A controls the maximum thickness of the soil at the groove to 0.8 cm. There is another type of grooving tool in use which does not control the thickness. The American Society for Testing Materials (III–3) and the American Association of State

through the sample along the symmetrical axis of the cup, holding the tool perpendicular to the cup at the point of contact (see Figs. III–1 and III–2).

FIGURE III–2. Liquid limit groove.

3. Turn the crank at a rate of about two revolutions per second, and count the blows necessary to close the groove in the soil for a distance of ½ in. (see

FIGURE III–3. Closed liquid limit groove.

Fig. III–3). The groove should be closed by a flow of the soil and not by slippage between the soil and the cup.

Highway Officials (III–2) both employ this second type of tool. The ASTM specifies that the depth of soil at the groove be about ⅜ in. (0.95 cm) and the AASHO specifies 1.0 cm.

4. Mix the sample in the cup and repeat steps 2 and 3 until the number of blows required to close the gap is substantially the same. (A difference of two or three blows probably indicates poor mixing of the sample.)

5. When a consistent value in the range of ten to forty blows has been obtained, take approximately 10 g of soil from near the closed groove for a water content determination.

6. By altering [14] the water content of the soil and repeating steps 2–5, obtain four water content determinations in the range of ten to forty blows.

7. Make a plot of water content against log of blows. Such a plot, known as a "flow curve" (see Numerical Example), is usually approximately a straight line.

II. Plastic Limit Determination

1. Mix thoroughly about 15 g of the moist soil.

2. Roll the soil on a glass plate with the hand [15] until it is ⅛ in. in diameter (see Fig. III–4).

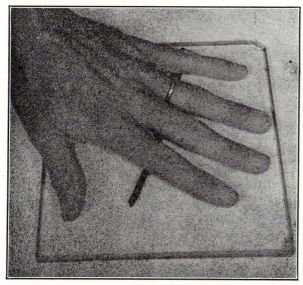

FIGURE III–4. Plastic limit determination.

3. Repeat step 2 until a ⅛ in. diameter thread shows signs of crumbling. (Figure III–5 shows three specimens. The one at the left is wetter than the

plastic limit, the center one at the plastic limit, and the right one drier than the plastic limit.)

4. Take some of the crumbling material obtained in step 3 for a water content determination.

5. Repeat steps 2–4 to obtain three determinations which can be averaged to give the plastic limit.

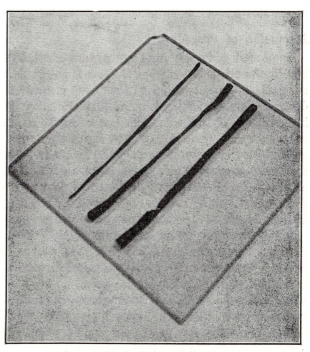

FIGURE III–5. Plastic limit threads.

III. Shrinkage Limit Determination [16]

1. Weigh a pat [17] of dry soil soon after it has been removed from a desiccator (it picks up moisture from the air).

2. Place a small dish in a larger one and fill the small one to overflowing with mercury. Cover the dish with a glass plate with prongs in such a way that the plate is flush with the top of the dish and no air is entrapped.

3. Wipe the outside of the small dish to remove adhering mercury, then place it in another large dish which is clean and empty.

4. Place the soil pat on the mercury and submerge it with the pronged glass plate, which is again made flush with the top of the dish (see Fig. III–6).

5. Weigh the mercury that is displaced by the soil pat.

---

[14] For some soils it has been found more convenient to start with the soil drier than the liquid limit and obtain values by increasing the water content. This is generally the quicker method for very fine-grained clays.

[15] It has been found that if a very small sample of soil is employed or if the thread is rolled with the tip of a finger, a different value for the plastic limit can be obtained than is obtained by the generally accepted method presented above. In order to eliminate such difficulties, consideration is being given to a mechanical device for the plastic limit test.

[16] The calculation of the shrinkage limit by the method proposed here requires the knowledge of the specific gravity of the soil.

[17] It is important that the pat be smooth and have rounded edges so that air will not be entrapped when the pat is immersed in the mercury.

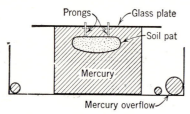

FIGURE III–6. Shrinkage limit determination.

Discussion of Procedure

The American Society for Testing Materials (III–3) and the American Association of State Highway Officials (III–2) permit the use of the "hand method" of determining the liquid limit in addition to that given in this chapter. In the hand method a tool is used to cut a groove (same size groove as in the mechanical method—see footnote 13 on page 24) in a soil paste contained in an evaporating dish. The water content is adjusted until the groove is closed by jarring the dish lightly against the heel of the hand ten times. The liquid limit is this adjusted water content.

A simplified procedure for determining the liquid limit has been investigated (III–16). The method is based on the assumption that the slope of the plot of blows on log scale against water content on log scale is a straight line with a constant slope. If this assumption were correct, the liquid limit could be obtained from one point on the curve. Based on 767 liquid limit tests, the discovery was made that for the soils investigated the liquid limit, w_1, could be found from [18]

$$w_1 = w_N \left(\frac{N}{25}\right)^{0.121}$$

in which w_N = the water content of the soil which closes in N blows in the standard liquid limit device.

The soil used for liquid and plastic limit determinations should not be dried prior to testing because drying may alter soil by causing the particles to subdivide or agglomerate, by driving off adsorbed water which is not completely regained on rewetting, or by effecting a chemical change [19] in any organic matter in the soil. These effects can significantly change [20] the limits, especially the liquid limit. The liquid limit of oven-dried organic soils tends to be lower than that of the undried. On the other hand, the effect of drying on some clays (particularly those composed of

minerals of the montmorillonite group) cannot be predicted. For example (III–8), the liquid limit of an air-dried sample of a soil was 20% greater than that determined on the soil not previously dried, whereas the liquid limit of an oven-dried sample of the same soil was 24% less than that determined on the undried soil.

In running the tests for the limits, we should use distilled water in order to minimize the possibility of ion exchange between the soil and any impurities in the water. Although the chances of such ion exchange are small, the added trouble of using distilled water is little enough to justify it.

Calculations

The limits, expressed as water contents, are obtained as follows.

I. *Liquid limit*, w_l, is read from the flow curve as the water content at twenty-five blows.[21]

II. *Plastic limit*, w_p, is the water content of the soil which crumbled when $\frac{1}{8}$-in. threads were rolled (the average of the consistent determinations made).

III. *Shrinkage limit*, w_s, is found from

$$w_s = \frac{\gamma_w V}{W_s} - \frac{G_T}{G_s} \qquad \text{(III–1)}$$

in which γ_w = unit weight of water,

W_s = weight of dry soil pat,

V = volume of dry soil pat

$$= \frac{\left\{\begin{array}{c}\text{weight of displaced mercury}\\ \text{(step 5, p. 25)}\end{array}\right\}}{13.55},$$

G_T = specific gravity of water at temperature of test (see Table A–2, p. 147, Appendix A),

G_s = specific gravity of soil grains.

IV. *Water-plasticity ratio*, B, is found from the equation [22]

$$B = \frac{w_n - w_p}{w_l - w_p} \qquad \text{(III–2)}$$

in which w_n = natural water content.

V. *Atterberg indices* can be calculated from the limits as follows:

(a) Plasticity index, I_p = liquid limit − plastic limit

(b) Flow index, I_f = slope of flow curve (see p. 28)

(c) Toughness index, $I_t = \dfrac{\text{Plasticity index}}{\text{Flow index}}$

[18] In reference III–16 a nomograph is furnished for the solution of the equation.

[19] See footnote on page 11.

[20] References III–6 and III–8 and unpublished work at M.I.T. Also see Fig. III–7.

[21] The liquid limit is occasionally taken as the water content at ten blows (III–12).

[22] Equation III–2 is the definition of B.

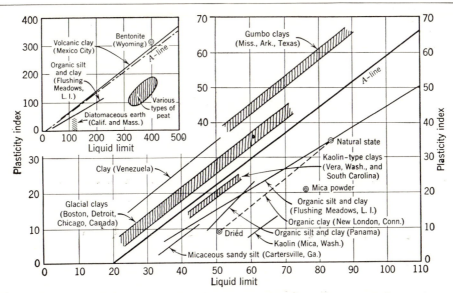

FIGURE III–7. Relation between liquid limit and plasticity index for typical soils. (From reference III–7.)

Results

Method of Presentation. The results of limit tests can be presented in a table like Table III–2. Often the limits are plotted against depth for a boring (see Fig. XII–8).

Typical Values. The limits may vary from very high values for extremely plastic clays to low values for the coarser-grained soils. We can obtain a good indication of typical limit values by studying Fig. III–7 (reference III–7). Figure XII–8, page 118, shows limits, plasticity index, and water plasticity ratio plotted against depth for a glacial soil.

In Table III–2 are listed limits for four soils which are quite different in their origin and composition. These values help illustrate the range that may be obtained.

TABLE III–2

Soil	w_l, %	I_p, %	w_s, %
Mexico City clay (volcanic origin, composed mostly of montmorillonite)	388.0	162.0	43.0
Boston blue clay (marine deposit of glacial clay composed partially of illite)	41.0	16.0	18.7
Morganza Louisiana clay (fluvial deposit, composed mostly of illite)	104.0	29.2	13.7
Beverly clayey silt (glacial deposit, composed of illite, limonite, and quartz)	19.5	3.2	13.3

Numerical Example

In the example on page 28 are presented the limits determinations on a glacial clay.

REFERENCES

1. Albin, Pedro, Jr., "Laboratory Investigation of the Variable Nature of the Blue Clay Layer below Boston," Master of Science Thesis, Massachusetts Institute of Technology, May, 1947.

2. American Association of State Highway Officials, *Standard Specifications for Highway Materials,* 1950.

3. American Society for Testing Materials, "Procedures for Testing Soils," Philadelphia, Pa., July, 1950.

4. American Society for Testing Materials, *Non-Metallic Materials,* Part II, *Constructional,* Philadelphia, Pa., 1946.

5. Atterberg, A., "Über die Physikalische Bodenuntersuchung, und über die Plastizitat der Tone," *Internationale Mitteilungen fur Bodenkunde,* Vol. 1, 1911.

6. Casagrande, A., "Research on the Atterberg Limits of Soils," *Public Roads,* Vol. 13, No. 8, October, 1932.

7. Casagrande, A., "Classification and Identification of Soils," *Transactions of the American Society of Civil Engineers,* Vol. 113, p. 901, 1948.

8. Casagrande, A., "Notes on Swelling Characteristics of Clay-Shales," Harvard University, Cambridge, Mass., July, 1949.

9. Casagrande, A., and R. E. Fadum, "Application of Soil Mechanics in Designing Building Foundations," *Transactions of the American Society of Civil Engineers,* Vol. 109, p. 383, 1944.

10. Civil Aeronautics Administration, "Airport Paving," U. S. Department of Commerce, May, 1948.

11. Hansen, J. Brinch, "Vane Tests in a Norwegian Quick-clay," *Géotechnique,* Vol. II, No. 1, June, 1950.

12. Housel, William S., *Laboratory Manual of Soil Testing Procedures,* University of Michigan, 1950.

13. Hyzer, Peter C., "An Analysis of Methods of Design of Flexible Pavements," Master of Science Thesis, Massachusetts Institute of Technology, May, 1949.

14. Skempton, A. W., and A. W. Bishop, "The Measurement of the Shear Strength of Soils," *Géotechnique,* Vol. II, No. 2, December, 1950.

15. War Department, "Soil Testing Set No. 1 and Expedient Tests," *War Department Technical Bulletin* TB5–253–1, June, 1945.

16. Waterways Experiment Station, "Simplification of the Liquid Limit Test Procedure," *Technical Memorandum* No. 3–286, June, 1949.

SOIL MECHANICS LABORATORY

ATTERBERG LIMITS

SOIL SAMPLE _Silty clay; gray at natural water content; inorganic; glacial origin; sedimentary deposit; extremely sensitive; medium plastic; very soft when remolded._

TEST NO. _L-33_

DATE _Nov. 20, 1949_

LOCATION _Union Falls, Maine_
BORING NO. _GC_   **SAMPLE DEPTH** _16.83_
SAMPLE NO. _GC-1-16.83_
SPECIFIC GRAVITY, G_s, _2.80_

TESTED BY _WCS_

PLASTIC LIMIT

DETERMINATION NO.	1	2	3
CONTAINER NO.	F-11	F-4	
WT. CONTAINER + WET SOIL IN g	22.116	21.844	
WT. CONTAINER + DRY SOIL IN g	20.419	20.187	
WT. WATER, W_w, IN g	1.697	1.657	
WT. CONTAINER IN g	13.069	13.178	
WT. DRY SOIL, W_s, IN g	7.350	7.009	
WATER CONTENT, w, IN %	23.1	23.6	

NATURAL WATER CONTENT

	1	2	3
	E-11	E-15	E-8
	17.527	16.971	17.356
	14.837	14.356	14.654
	2.690	2.615	2.702
	7.835	7.503	7.553
	7 002	6.853	7.101
	38.4	38.2	38.0

LIQUID LIMIT

DETERMINATION NO.	1	2	3	4	5
NO. OF BLOWS	29	21	17	13	
CONTAINER NO.	F-17	F-16	F-8	F-29	
WT. CONTAINER + WET SOIL IN g	22.244	21.190	21.268	26.115	
WT. CONTAINER + DRY SOIL IN g	19.443	18.781	18.746	22.102	
WT. WATER, W_w, IN g	2.801	2.409	2.522	4.013	
WT. CONTAINER IN g	12.737	13.241	13.058	13.288	
WT. DRY SOIL, W_s, IN g	6.706	5.540	5.688	8.814	
WATER CONTENT, w, IN %	41.8	43.5	44.4	45.5	

WATER - PLASTICITY RATIO, $B = \dfrac{w_n - w_p}{w_l - w_p}$

SHRINKAGE LIMIT

DETERMINATION NO.	1	2
UNDISTUBED OR REMOLDED SOIL PAT	REMOLDED	
WT. DRY SOIL PAT, W_s, IN g	19.66	
WT. CONTAINER + HG. IN g	257.32	
WT. CONTAINER IN g	109.35	
WT. HG. IN g	147.97	
VOL. SOIL PAT, V, IN cc	10.88	
SHRINKAGE LIMIT, w_s, IN %	19.3	

$$w_s = \frac{\gamma_w V}{W_s} - \frac{G_I}{G_s}$$

FLOW CURVE

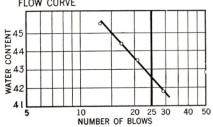

RESULT SUMMARY

PLASTIC LIMIT	NATURAL WATER CONTENT	LIQUID LIMIT	SHRINKAGE LIMIT	B VALUE	PLASTICITY INDEX	FLOW INDEX	TOUGHNESS INDEX
23.4	38.2	42.6	19.3	77.0	19.2	10.9	1.76

REMARKS·

CHAPTER

IV

Grain Size Analysis

Introduction

In the early days of soil mechanics many people thought that the size of the individual particles of a soil would prove to be its most important characteristic. The ensuing years of research and experience have failed to substantiate this belief. Partly because of inertia resulting from this early belief and partly because a knowledge of the grain size distribution of a coarse soil can be very useful, most laboratories of soil mechanics run grain size analyses as a routine test on almost every soil that comes to them.

In this chapter, the three general procedures of analysis presented are sieve, hydrometer, and combined analyses. A sieve analysis consists of shaking the soil through a stack of wire screens with openings of known sizes; the definition of particle diameter for a sieve test is, therefore, the side dimension of a square hole. The hydrometer method is based on Stokes' equation for the velocity of a freely falling sphere; the definition of particle diameter for a hydrometer test is, therefore, the diameter of a sphere of the same density which falls at the same velocity as the particle in question. The combined analysis employs both the sieve and the hydrometer tests; thus the definition of particle size is the size of a square opening for the larger grains and the diameter of the equivalent sphere for the smaller soil particles.

The test procedure which should be followed depends on the soil in question. If nearly all its grains are so large that they cannot pass through square openings of 0.074 mm (No. 200 screen), the sieve analysis is preferable. For those soils which are nearly all finer than a No. 200 screen, the hydrometer

test is recommended.[1] For silts, silty clays, etc., which have a measurable portion of their grains both coarser and finer than a No. 200 sieve, the combined analysis is needed.

There are many reasons, both practical and theoretical, why the grain size distribution curve of a soil is only approximate. Some of these reasons are discussed later in this chapter, and others are discussed in reference IV–1. The accuracy of distribution curves for fine-grained soils is more questionable than the accuracy of the curves for the coarser soils. In the case of residual soils the term "individual particle" is rather arbitrary since the particle size depends on the degree of disaggregation that is imparted mechanically prior to testing. The significance of grain size data for such soils is extremely doubtful. And even if a grain size curve which was exact could be obtained, it might be of only limited value. Although the behavior of cohesionless soils can often be related to particle size, as pointed out below, the behavior of cohesive soils depends more on such things as type of clay mineral[2] and geological history than on particle size. The limits described in Chapter III give much more information as to the behavior of clays than the grain size data.

[1] For particles finer than approximately 0.0002 mm, the use of the hydrometer test is in question. See footnote 16, page 34. A centrifuge or supercentrifuge can be used in conjunction with Stokes' equation for size analyses of grains smaller than 0.0002 mm.

[2] Sometimes we can obtain indications of the type of clay mineral from a grain size distribution curve. Usually a method of analysis more complicated than the hydrometer test is required if the curve is to be carried to diameters small enough to be useful for such identification.

In spite of their serious limitations, grain size curves, particularly those of sands and silts, do have practical value. Both theory and laboratory experiments show that soil permeability (the ease with which a fluid will flow through the soil) and capillarity (the attraction or the retention of water above the water table) are related to an effective particle diameter. The permeability of a cohesionless soil is approximately proportional to the square of an effective diameter. The diameter read from a grain size curve at the "10% finer" point is sometimes used as the effective diameter (IV–6). The rise of water in a capillary opening is proportional to the reciprocal of the diameter of the opening. If pore size can be related to particle size, a relationship between capillary rise and particle size can be obtained.

The method of designing inverted filters (IV–10) for dams, levees, etc., uses the particle size distribution of the soils involved. This method is based on the relationship of grain size to permeability, along with experimental data on the grain size distribution required to prevent the migration of particles when water flows through the soil. Also the present criterion for establishing susceptibility of soils to frost damage is based on grain size.

Grain size curves have been widely used in the identification and classification of soils. At the top of Fig. IV–8 is plotted the M.I.T. grain size classification, which is one of several such classifications available. Often the percentage of a soil's weight finer than a certain size is studied in connection with other properties in the hope that some relationship between them may be found. For example, in Fig. XII–8, the percentage of a particular soil finer than 0.005 mm has been plotted against depth, along with strength and other properties. For this soil, the weaker samples seem to have a smaller percentage finer than 0.005 mm.

From the preceding discussion, we see that grain size curves of coarser-grained soils are often very useful, but that those of fine-grained soils may have quite limited application.

SIEVE ANALYSIS

Apparatus and Supplies

Special
   1. Set of sieves

General
   1. Brush (for cleaning sieves)
   2. Beam balance (0.1 g sensitivity)
   3. Drying oven
   4. Desiccator[3]

[3] The desiccator may not be required. See page 10.

   5. Syringe
   6. Large pan
   7. Mortar and rubber-tipped pestle

Figure IV–1 shows a set of sieves which can be nested. The selection[4] of sieves for a given test depends on the soil to be tested, e.g., the coarser the

FIGURE IV–1. Sieves for grain size analysis. (From reference IV–9.)

soil, the larger the top sieve should be. A good spacing of soil diameters on the grain size distribution curve (see Fig. IV–8) will be obtained if a nest of sieves is used in which each sieve has an opening ap-

[4] For soil finer than a No. 10 sieve (2 mm), the ASTM requires the following sieves: Nos. 20, 40, 60, 140, and 200. The AASHO specifies Nos. 40, 60, and 200.

proximately one-half that of the coarser sieve above it in the nest.

Table IV–1 is a partial list of sieve sizes in current use. The Tyler series are specified by "mesh," which is the number of openings per inch of screen. The

TABLE IV–1

SIEVE SERIES *

| | Tyler Standard | | Wire Diameter, | | U. S. Bureau of Standards | |
| | Opening | | | | Opening | |
Mesh	in.	mm	in.	Number	in.	mm
..	3.0	76.2	0.207	..	4.00	101.6
..	2.0	50.8	0.192	..	2.00	50.8
..	1.050	26.67	0.148	..	1.00	25.4
..	0.742	18.85	0.135	..	0.750	19.1
..	0.525	13.33	0.105	..	0.500	12.7
..	0.371	9.423	0.092	..	0.375	9.52
3	0.263	6.680	0.070	3	0.250	6.35
4	0.185	4.699	0.065	4	0.187	4.76
6	0.131	3.327	0.036	6	0.132	3.36
8	0.093	2.362	0.032	8	0.0937	2.38
9	0.078	1.981	0.033	10	0.0787	2.00
10	0.065	1.651	0.035	12	0.0661	1.68
14	0.046	1.168	0.025	16	0.0469	1.19
20	0.0328	0.833	0.0172	20	0.0331	0.840
28	0.0232	0.589	0.0125	30	0.0232	0.590
35	0.0164	0.417	0.0122	40	0.0165	0.420
48	0.0116	0.295	0.0092	50	0.0117	0.297
60	0.0097	0.246	0.0070	60	0.0098	0.250
65	0.0082	0.208	0.0072	70	0.0083	0.210
100	0.0058	0.147	0.0042	100	0.0059	0.149
150	0.0041	0.104	0.0026	140	0.0041	0.105
200	0.0029	0.074	0.0021	200	0.0029	0.074
270	0.0021	0.053	0.0016	270	0.0021	0.053
400	0.0015	0.038	0.001	400	0.0015	0.037

* Data from reference IV–9.

list of Tyler sieves in Table IV–1 (except for the top two and the 9 and 60 mesh) are selected so that the opening in each sieve is $\sqrt{2}$ times that of the sieve listed immediately below it. The sieves of the U. S. Bureau of Standards series are numbered; the numbers are based on the size of opening. Thus a No. 100 sieve has openings which are twice as large as those of a No. 200 sieve. Specifications usually permit variations in average openings ranging from 2% for the coarse sieves to 7% for the fine sieves.

Recommended Procedure

1. Weigh to 0.1 g each sieve which is to be used. Make sure each sieve is clean before weighing it.

2. Select with care a test sample which is representative of the soil to be tested; break the soil into its individual particles with the fingers or a rubber-tipped pestle.

3. Weigh to 0.1 g a specimen of approximately 500 g of oven-dried soil. If the soil to be tested has

many particles coarser than the openings in a No. 4 sieve, a larger weight of soil should be used.

4. Sieve the soil through a nest of sieves by hand shaking, using a motion of horizontal rotations or using a mechanical shaker, if available. At least 10 minutes of hand sieving is desirable for soils with small particles.

5. Weigh to 0.1 g each sieve and the pan with the soil retained on them.

6. Subtract the weights obtained in step 1 from those of step 5 to give the weight of soil retained on each sieve. (The sum of these retained weights should be checked against the original soil weight.)

7. If a sizable portion of soil is retained on the No. 200 sieve, it should be washed. This is done by placing the sieve and retained soil in a pan and pouring clean water on the screen. Use a spoon or glass rod to stir the slurry.[5] Recover the soil which is washed through; dry and weigh it. The weight of soil recovered should be subtracted from the weight retained on the No. 200 sieve and added to the weight retained in the pan as determined in step 6.

Discussion of Procedure

The method of weighing the sieve plus soil rather than attempting to remove the soil from the sieve for weighing is suggested because it has been found that soil is often lost during the removing. Even using this suggested procedure, we have to be careful to minimize the loss of soil during the sieving.

Step 4 recommends that the sieving consist of approximately 10 minutes of horizontal shaking. A horizontal motion was suggested instead of a vertical one since it has been found more efficient (IV–2) and since less soil escapes from the nest of sieves during horizontal shaking. The amount of shaking required depends on the shape and number of particles. As an example of the fact that the shaking time required is increased as the number of particles is increased, for crushed quartz it was found (IV–2) that, in a given time, the percentage passing was 25% less for a 250-g sample than it was for a 25-g sample. Since a given weight of a fine-grained soil contains more particles than an equal weight of a coarse-grained one, more shaking time is necessary for the finer-grained soils.

Calculations

(1) Percentage retained on any sieve

$$= \frac{\text{wt. of soil retained}}{\text{total soil wt.}} \times 100\%$$

[5] In case the lumps were not completely broken down into their individual particles in step 2, allow the slurry to sit until they can be.

(2) Cumulative percentage retained on any sieve = sum of percentages retained on all coarser sieves

(3) Percentage finer than any sieve size = 100% − cumulative percentage retained

Figure IV–2a shows a type of hydrometer [6] which has proved to be satisfactory for soil testing. It can be obtained in three specific gravity ranges: namely, 0.995 to 1.030, 0.995 to 1.040, 1.000 to 1.060. The first

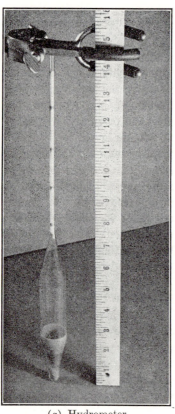

(a) Hydrometer

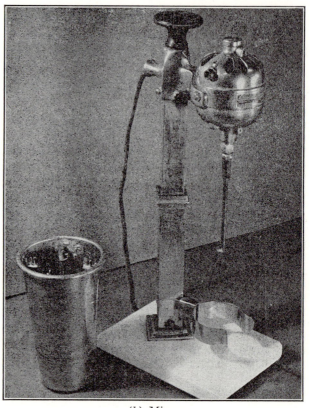

(b) Mixer

FIGURE IV–2

HYDROMETER ANALYSIS

Apparatus and Supplies

Special

1. Hydrometer
2. Mixer
3. Deflocculating agent
4. Constant-temperature bath (optional)

General

1. Two graduated cylinders (1-liter capacity)
2. Distilled water supply
3. Balance (0.1 g sensitivity)
4. Drying oven
5. Desiccator
6. Thermometer (graduated to 0.1° C)
7. Syringe
8. Large evaporating dishes
9. Spatula
10. Timer

of these ranges is preferable because the larger space between 0.001 division makes accuracy easier to obtain. Also, the maximum hydrometer reading is almost always less than 1.030.[7] The mixer shown in Fig. IV–2b is a common electric mixer; the cup has fins down the inside to aid mixing.

A constant-temperature bath is desirable for long-duration tests, especially in laboratories which experience large fluctuations of temperature.

Recommended Procedure

Before running his first hydrometer analysis, a student should practice placing the hydrometer in the

[6] The type of hydrometer shown in Fig. IV–2a can be purchased from Eimer and Amend Company, 635 Greenwich Street, New York 14, New York.

[7] As pointed out on page 34, the maximum soil weight used for the hydrometer analysis should be approximately 50 g. The substitution of 50 g for W_s, together with reasonable soil properties in Eq. IV–2, will give a maximum hydrometer reading near 1.030.

suspension and reading it (see footnote 15). To insert the hydrometer correctly, hold it carefully by the stem with both hands (see Fig. IV–3) and lower it slowly to the depth at which it floats. The inserting process should take about 5 seconds. A hydrometer that is properly inserted will neither bob nor rotate appreciably when released.

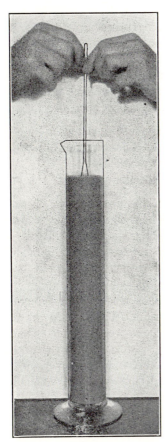

Figure IV–3. Inserting an hydrometer.

A calibration for the hydrometer used will be needed. See page 156, Appendix B, for an outline of the calibration procedure.

1. Mix a moist [8] specimen of soil, representing approximately 50 g [9] dry weight, with distilled water to form a smooth thin paste.

2. Add a deflocculating agent [10] to the paste and wash the mixture into the cup of the mixing machine by using a syringe.

---

[8] The complete drying of a clay may cause the individual particles to change size or undergo some other form of permanent alteration.

[9] If the soil contains many larger particles which settle rapidly, a larger weight of soil may be used.

[10] A deflocculating agent can either act as a protective colloid on the soil particle or alter the electrical charge on the particle to prevent the formation of soil flocs. Two commonly

3. Mix the suspension in the machine [11] until the soil is broken down into its individual particles (approximately 10 minutes).

4. While the soil and water are being mixed, fill a graduated jar with distilled water. Use this jar of water to store the hydrometer in between readings.

5. After mixing, wash the specimen into a graduated cylinder and add enough distilled water to bring the level to the 1000-cc mark.

6. Mix the soil and water in the graduate by placing the palm of the hand over the open end and turning the graduate upside down and back (see Fig. IV–4). When the graduate is upside down (Fig. IV–4b) be sure no soil is stuck to the base of the graduate.

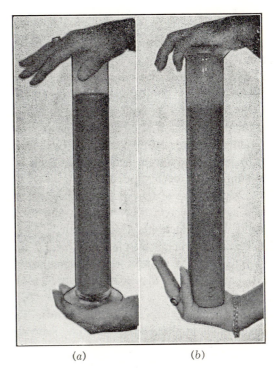

(a)   (b)

Figure IV–4. Mixing a jar of suspension.

7. After shaking it for approximately 30 seconds, replace the graduate on the table, insert the hydrometer in the suspension, and start the timer.

used deflocculants are water glass (sodium silicate) and Daxad No. 23 (polymerized sodium salts of substituted benzoid alkyl sulfonic acid), made by the Dewey and Almy Chemical Company of Cambridge, Mass. The easiest way to select the best deflocculant and the amount to use for a given soil is by trial. Two cc of a freshly prepared 10% solution of Daxad No. 23 has been found satisfactory for many clays; likewise, one-half to one cc of 40° Baumé sodium silicate has been found satisfactory for many clays. Sodium pyrophosphate has also been used successfully.

[11] The soil is sometimes dispersed by air instead of by mechanical mixing.

8. Take hydrometer readings at total elapsed times [12] of $\frac{1}{4}$, $\frac{1}{2}$, and 1 and 2 minutes without removing the hydrometer. The suspension should be remixed, and this set of four readings repeated until a consistent pair of sets have been obtained.

9. After the 2-minute reading, remove the hydrometer, remix, and restart the test, but take no reading until the 2-minute one. For this reading and all the following ones, insert the hydrometer just before reading. Before each insertion of the hydrometer, dry the stem. [13]

10. Take hydrometer readings at total elapsed time intervals of 2, 5, 10, 20 minutes, etc., approximately doubling the previous time interval. The hydrometer should be removed from the suspension and stored in the graduate of distilled water after each reading. Take frequent temperature measurements of the suspension.

11. Take temperature observations and hydrometer readings in the jar of distilled water every 20 or 30 minutes; add warm or cool water to keep it at the same temperature as that of the suspension. Attempt to minimize [14] temperature variations by keeping the test graduates away from heat sources such as radiators, sunlight, or open windows.

12. Keep the top of the jar containing the soil suspension covered to retard evaporation and to prevent the collection of dust, etc., from the air.

13. Obtain the height of meniscus rise of pure distilled water on the stem of the hydrometer. [15] This height, known as the meniscus correction, is used in the calculations.

14. Continue taking observations until the hydrometer reads approximately one, i.e., around 1.001, or until readings have been obtained at elapsed times large enough to give the minimum soil particle diameter desired (see footnote 1, page 29).

15. After the final reading, pour the suspension into large evaporating dishes; take unusual care to avoid losing any soil.

16. Evaporate the suspension to dryness in the oven, cool the dishes in the desiccator, and weigh to 0.1 g.

17. Clean the dishes and weigh them. The weight of the dishes subtracted from that determined in step 16 gives the weight of dry soil used.

Discussion of Procedure

The preceding procedure depends, as do the other wet methods of grain size analysis, on Stokes' equation for the terminal velocity of a falling sphere. There are a number of assumptions [16] in Stokes' equation which are not completely fulfilled in the hydrometer method. Among them are:

1. No interference of particles by other particles or by the walls of the container.
2. Spherical particles.
3. Known specific gravity of the particles.

The first of the above assumptions can be practically satisfied by limiting the maximum concentration of soil in the suspension. Research (IV–11) has shown that if no more than 50 g of dry soil are used in 1000 cc of suspension, the effects of interference are negligible.

The shapes of most of the largest particles (i.e., larger than about 0.005 mm) in the hydrometer analysis can be represented by spheres with reasonable accuracy. Most soil particles smaller than approximately 0.005 mm, however, are shaped like a plate, the length or width of which is five to three hundred times its thickness. The plate shape which most clay particles possess is well illustrated by the electron photomicrograph of a kaolinite clay particle shown in Fig. IV–5. The diameter of the sphere which would fall through water at the same velocity as a plate, such as that in Fig. IV–5, would be smaller than the length or width of the plate because a plate shape is not the one which receives the minimum resistance to fall. The fall of a plate through water is somewhat like the downward drifting of a leaf from a tree.

As pointed out in the previous chapters, the soil particle in suspension is surrounded by a water film. There is still question as to the thickness [17] of water

---

[12] The elapsed times suggested are merely convenient ones; the calculations can be made for any times.

[13] Any water adhering to the stem above the point where the surface of the suspension intersects the stem will make the hydrometer too heavy, and thus cause readings which are too small.

[14] The ASTM and AASHO require that the test graduate be kept in a constant-temperature bath. Such a bath is desirable.

[15] The stem should be kept very clean since any foreign matter on it may alter the height of meniscus rise. A slightly larger meniscus rise may be obtained if the hydrometer rises to its steady position rather than sinks to it.

[16] Stokes' equation neglects the kinetic motion of the molecules of the suspending medium. For particles fine enough this kinetic motion can offset the effect of gravity on the soil particle; such a particle is in Brownian motion. For particles below 0.0002 mm in diameter the validity of Stokes' equation for the hydrometer analysis is in question.

[17] Water film thicknesses varying from 13 Å to over 1000 Å have been reported (IV–3). The thickness of water films on soil particles have not yet been measured directly; the thick-

films on clay particles. However, film thicknesses of the order of magnitude of the particle thickness have been suggested (IV–4). Since the volume of such a particle would be approximately two-thirds water,[18] its specific gravity would be nearer 1.8 than the 2.7 or 2.8 determined by the procedure given in Chapter II of this book. The use of a specific gravity which

The phenomena mentioned above combine their effects; the first two tend to make the diameter computed by the hydrometer procedure too small and the third too large. The net effect is that the hydrometer analysis indicates diameters which are smaller than the length or width of the plate-shaped particle. This reasoning is supported by experimental evidence.[19]

FIGURE IV–5. Electron photomicrograph of a kaolinite particle. (Courtesy of C. E. Hall of M.I.T. From reference IV–3.)

is too large in Stokes' equation gives diameters which are too small. On the other hand, the adsorbed water film makes the actual particle settling out of the suspension larger than the mineral grain.

In addition to the limitations to the use of Stokes' equation for particle size analysis, there are many approximations made in the development of the hydrometer method. These approximations are covered

---

nesses to date have been obtained by procedures employing various assumptions. The current feeling seems to favor the lower side of the above range. The thicknesses appear to depend on the type of mineral, the exchangeable ions, and the pressure acting on the films. In line with the above information, a film thickness approaching the particle thickness appears to be too large for a kaolinite but not unreasonable for a montmorillonite.

[18] To be consistent, the weight of "dry" soil used in Eq. IV–2 should be the mineral grain plus water film. However, the

determination of such a weight appears to be infeasible. As pointed out in Chapter I, drying to 105° C removes some but not all of the soil water. It can be seen that the grain size curve for some soils depends on the method employed to get the "dry" soil weight. For an extreme example, if the Mexico City clay whose drying curve is shown in Fig. I–6 were dried at 190° C instead of 105° C, the values of percentage finer from Eq. IV–2 would be approximately 7% larger.

[19] Hofmann (IV–7) compared the size of particles determined by the electron microscope and by Stokes' equation for both

completely elsewhere (IV–1); their effects are less important than the other considerations presented in this discussion.

There is another point that must be considered. This is the change in the particle size which may occur when the soil is put in suspension. Grim (IV–5) has pointed out that montmorillonite mineral particles and some illite mineral particles break down into smaller ones when stirred into suspension in water. This subdividing, in addition to changes which may occur in the thickness of water films [20] when the clay is put into suspension, makes it difficult, if not impossible, to obtain by the hydrometer method the particle sizes of some soils as they exist in nature. However, this does not subtract from the value of a hydrometer analysis for soil identification or classification.

Calculations (See Alternate Methods of Calculation on page 38)

(a) The effective diameter, D, can be computed from

$$D = \sqrt{\frac{18\mu}{\gamma_s - \gamma_w}} \sqrt{\frac{Z_r}{t}} \qquad (IV\text{--}1)$$

in which μ = viscosity of water at the temperature [21] of the test (see Table A–3, p. 148),

γ_s = unit weight of the soil grains,

γ_w = unit weight of water at the temperature [21] of the test (see Table A–2, p. 147),

Z_r = distance from surface of suspension to the center of volume of the hydrometer,[22]

t = total elapsed time.

_____

quartz and kaolinite. He found that Stokes' equation applied for quartz but that the length and width of the kaolinite particle were larger than the equivalent diameter obtained from Stokes' equation. Hofmann's findings are substantiated by East (IV–3). The face of the particle shown in Fig. IV–5 has the same area as a disk of diameter 1.3 μ (it has a thickness of 0.083 μ, which is approximately 115 unit cells). The particle was obtained from a fraction having equivalent diameters from 0.4 μ to 0.8 μ, as determined by a procedure based on Stokes' equation. Casagrande (IV–1) compared the equivalent diameter, as determined by the hydrometer method, with the particle size, as determined by a microscope, for both quartz and mica powder. Good agreement was obtained for the quartz; however, the equivalent diameter of the mica was less than one-third of the size measured with the microscope. The fact that the mica particles were not small enough to have water films which were sizable relative to the particle size may have caused their equivalent diameters to appear smaller in relation to the measured size than Hofmann's or East's clay particles.

[20] The deflocculating agent may change the thickness of the water film.

[21] The temperature of the test is the average temperature for the elapsed time t.

[22] Z_r is obtained from a hydrometer calibration curve (see

Figure IV–6 gives the term $\sqrt{\dfrac{18\mu}{\gamma_s - \gamma_w}}$ as a function of temperature for use in Eq. IV–1.

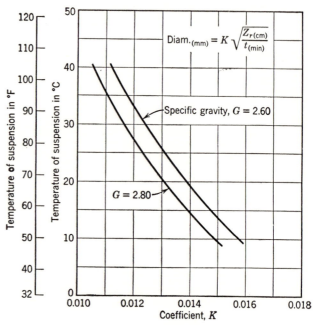

FIGURE IV–6. Chart for aid in solving Stokes' equation. (Courtesy of H. P. Aldrich of M.I.T.)

(b) The percentage finer, N, can be computed from

$$N = \frac{G}{G - 1} \frac{V}{W_s} \gamma_c (r - r_w) \times 100\% \qquad (IV\text{--}2)$$

in which G = specific gravity of solids,

V = volume of suspension (1000 cc),

W_s = weight of dry soil,

γ_c = unit weight of water at temperature (usually 20° C) of hydrometer calibration (see Table A–2, p. 147),

r = hydrometer reading in suspension,

r_w = hydrometer reading in water (at same temperature as suspension).

page 156, Appendix B). The hydrometer reading should be corrected for the meniscus rise on the hydrometer stem before entering a calibration curve. This means that the height of meniscus rise, as measured in water, must be numerically added to the hydrometer reading. Since the surface of the suspension rises when the hydrometer is inserted in it, a correction known as the immersion correction should also be made. This is done by subtracting half the volume of the hydrometer bulb divided by the cross-sectional area of the jar from the distance from the surface to the center of the bulb. A hydrometer calibration curve such as shown in Fig. B–7, page 156, should have two lines on it, one with and the other without the immersion correction.

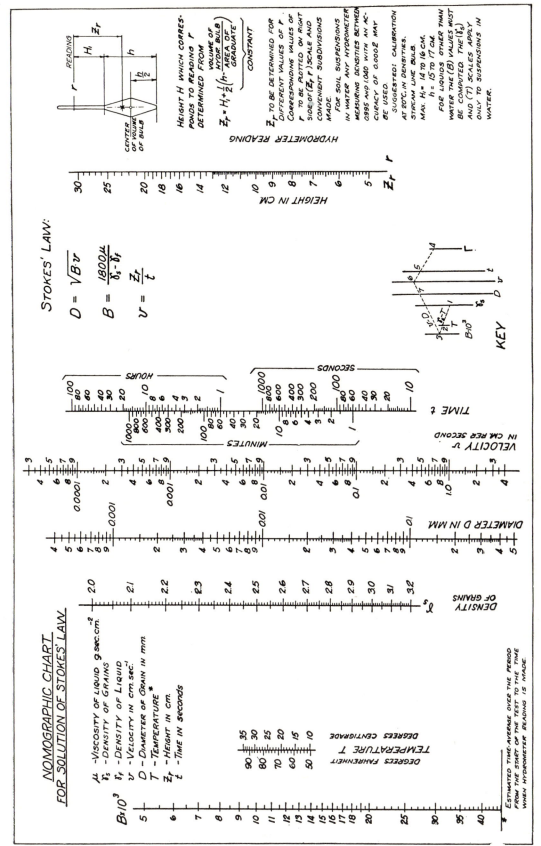

Figure IV-7. (Courtesy of A. Casagrande.)

Alternate Methods of Hydrometer Test Computations

Several types of nomographic charts for the solution of Stokes' equation and several types of cards for plotting grain size curves have been developed. One such chart is Fig. IV–7,[23] which was prepared by A. Casagrande. The use of Casagrande's nomograph is explained by the key and the notes on it. Nomographs are desirable for routine testing.

COMBINED ANALYSIS [24]

Recommended Procedure

The desirable procedure for a combined analysis depends to a certain extent on the soil to be tested. If the soil consists of only a small portion of fine material, as indicated by low dry strength (lumps of dry soil easily powdered by the fingers) or low plasticity, use Alternate a below; otherwise use Alternate b.

Alternate a

1. Oven-dry the soil and then break up all lumps with the fingers or a rubber-tipped pestle.

2. Run a sieve analysis as outlined on page 31; wash the soil retained on the No. 200 sieve. (See step 7, page 31.)

3. Weigh out, to 0.01 g, approximately 50 g of the dry soil retained in the pan [25] from the sieve analysis. Run a hydrometer analysis as outlined, starting on page 33, with two exceptions: (a) no deflocculant [26] is added and (b), since the dry soil weight was obtained at the start of the hydrometer test, there is no need to collect the suspension at the end of the test.

Alternate b

1. Add distilled water to the soil sample and work it into a slurry. Make sure that all lumps are broken up into individual particles. Use soil which has never been dried; the amount of soil should be enough to give approximately 350 g dry weight.

[23] Several changes in Casagrande's nomenclature have been made to make it agree with that used in the rest of this chapter.

[24] A student doing this test for the first time should be able to run a combined analysis on a silty soil in 2 to 3 hours and to do the computations in about 2 hours.

[25] A more representative sample can be obtained by using the soil washed through the No. 200 sieve in addition to that which was shaken through. Such a procedure requires more time since the soil washed through the sieve must be dried before the start of the hydrometer test or the suspension must be collected, dried, and weighed at the end of the test.

[26] The grains of a soil on which a combined analysis is run, such as a silt, are usually of large enough size not to flocculate. If flocculation is observed in the combined analysis, add a deflocculant to the suspension in the graduate.

2. Wash the slurry through a No. 200 sieve.

3. Oven-dry that soil retained on the No. 200 sieve, and run a sieve analysis as outlined on page 31.

4. Carefully wash the suspension, passing the No. 200 sieve into a graduated jar, and fill to the 1000-cc mark with distilled water.

5. Mix the suspension and take a ¼-minute reading. If this reading is so high that it is off the hydrometer stem, thoroughly mix the suspension, pour some off, and replace with water. In such a manner, select a sample size which will give a ¼-minute reading of approximately 1.030. Dry and weigh the soil in the suspension which was poured out.

6. Continue with the selected suspension as outlined in Hydrometer Analysis, starting with step 6. Prepare the graduated jar of distilled water required in step 4.

Discussion of Procedure

The discussions given following the sieve analysis and hydrometer analysis apply, of course, to the corresponding parts of the combined analysis. It should be re-emphasized that the two parts of the combined analysis are based on different definitions of particle size. As pointed out previously, the effective diameter of a plate-shaped clay particle is smaller than the size of sieve opening through which it would just pass. On the other hand, the effective diameters of the largest particles obtained by a hydrometer analysis are usually larger than their sizes as determined by sieve analysis. Many tests have shown that, in the overlap zone (that part of the curve which can be determined by both methods), the hydrometer analysis indicates larger particle sizes than the sieve analysis. The particles in this overlap zone are usually not plate shaped.

Calculations

Compute the particle size and percentage finer for the two parts of the combined analysis as shown in sieve analysis and hydrometer analysis. The weight of dry soil to be used in computing the sieve analysis should be the total sample. If alternate b was used, this total weight is obtained by adding the weight of soil used in the sieve analysis, plus that used in the hydrometer analysis, plus any poured off to bring the ¼-minute hydrometer reading to 1.030 or below. The corrected percentage, N', is found as follows for either of the alternate procedures of the combined analysis:

$$N' = N \cdot \frac{W_1}{W_s} = N \cdot \% \text{ finer than No. 200 sieve} \quad \text{(IV–3)}$$

in which N = percentage finer, computed as shown on page 36,

W_1 = weight of dry soil passing the No. 200 sieve,

W_s = total weight of dry soil used for sieve analysis computation.

Results

Method of Presentation. The results of a grain size analysis are usually presented in the form of a distribution curve (see Fig. IV–8); this curve is obtained by plotting particle diameter against per cent finer. The results can also be given by listing the percentage of the soil weight which falls into the different textural groups (see the Numerical Example below). For a boring, the percentage finer than a given size can be plotted against depth, as is done in Fig. XII–8.

Typical Values. The uniformity of a soil can be expressed by the uniformity coefficient, which is the ratio of D_{60} to D_{10}, where D_{60} is the soil diameter of which 60% of the soil weight is finer and D_{10} is the corresponding value at 10% finer. A soil having a uniformity coefficient smaller than about 2 would be considered "uniform."

There are several grain size classifications which give good indications of typical particle sizes. For example, by the M.I.T. classification, soil diameters between 0.002 and 0.06 mm are called silt, and diameters between 0.06 and 2.0 mm are called sand.

Numerical Example

On pages 40–42 is presented the combined analysis on a silty sand. Based on the M.I.T. classification, the sample consists of 2% gravel, 85% sand, 12% silt, and 1% clay. It has a uniformity coefficient of 10. Therefore, the soil would be termed a well-graded, silty sand.

REFERENCES

1. Casagrande, A., "The Hydrometer Method for Mechanical Analysis of Soils and Other Granular Materials," Cambridge, Mass., June, 1931.
2. DallaValle, J. M., *Micromeritics*, Pitman Publishing Corp., New York, 1948.
3. East, W. H., "Water Films in Monodisperse Kaolinite Fractions," Doctor of Science Thesis, Department of Metallurgy, Massachusetts Institute of Technology, August, 1949.
4. Endell, Kurd, "The Swelling Capacity of Clays in the Construction Foundation and Its Technical Significance," *Die Bautecknik*, Vol. 19, No. 19, May, 1941.
5. Grim, R. E., "Modern Concepts of Clay Minerals," *Journal of Geology*, Vol. L, No. 3, April–May, 1942.
6. Hazen, A., "Discussion of 'Dams on Sand Foundations' by A. C. Koenig," *Transactions of the American Society of Civil Engineers*, Vol. 73, p. 199, 1911.
7. Hofmann, Ulrich, "Recent Advances in the Chemistry of Clay," *Die Chemie*, Vol. 55, No. 37/38, September, 1942.
8. Taylor, D. W., *Fundamentals of Soil Mechanics*, John Wiley and Sons, New York, 1948.
9. Tyler, W. S., Co., *Catalog 53*, Cleveland, Ohio, 1947.
10. Waterways Experiment Station, "Field and Laboratory Investigation of Design Criteria for Drainage Wells," *Technical Memorandum* No. 195–1, Vicksburg, Miss., October, 1942.
11. Weatherly, W. C., "The Hydrometer Method for Determining the Grain Size Distribution Curve of Soils," Master of Science Thesis, Department of Civil Engineering, Massachusetts Institute of Technology, 1929.

SOIL MECHANICS LABORATORY

SIEVE ANALYSIS

SOIL SAMPLE _Silty Sand: grayish brown; well graded,_
subrounded particles; mostly quartz.

LOCATION _Hadley, N.Y._

BORING NO. _2_   SAMPLE DEPTH _10_

SAMPLE NO. _5_

SPECIFIC GRAVITY, G_s, _2.75_

TEST NO. _G-2 (S)_

DATE _Oct. 13, 1949_

TESTED BY _D D_

SOIL SAMPLE WEIGHT

CONTAINER NO. _A 5_

WT. CONTAINER + DRY SOIL IN g _913.4_

WT. CONTAINER IN g _488.7_

WT. DRY SOIL, W_s, IN g _424.7_

SIEVE NO.	SIEVE OPENING IN mm	WT. SIEVE IN g	WT. SIEVE + SOIL IN g	WT. SOIL RETAINED IN g	PERCENT RETAINED	CUMULATIVE PERCENT RETAINED	PERCENT FINER
4	4.76	521.5	521.5	0	0	0	100
8	2.38	390.0	390.0	0	0	0	100
20	.84	367.7	456.2	88.5	20.8	20.8	79.2
40	.42	367.0	444.9	77.9	18.4	39.2	60.8
100	.149	428.0	589.6	161.6	38.1	77.3	22.7
200	.074	300.4	334.4 -5.0*	29.0	6.8	84.1	15.9
Pan	—	335.9	398.5 +5.0*	67.6	15.9	100.0	—

REMARKS: * 5.0 g soil washed through #200 sieve.

SOIL MECHANICS LABORATORY

HYDROMETER ANALYSIS

SOIL SAMPLE _Silty Sand: grayish brown; well graded, subrounded particles; mostly quartz._

LOCATION _Hadley, N.Y._

BORING NO. _2_ SAMPLE DEPTH _10_

SAMPLE NO. _5_

SPECIFIC GRAVITY, G_s, _2.75_

SOIL SAMPLE WEIGHT

CONTAINER NO. _A8_

WT. CONTAINER + DRY SOIL IN g _449.2_

WT. CONTAINER IN g _400.8_

WT. DRY SOIL, W_s, IN g _48.4_

TEST NO. _G-2 (H)_

DATE _Oct. 14, 1949_

TESTED BY _D D_

HYDROMETER NO. _7365_

MENISCUS CORRECTION _0.5_

$$N = \frac{G}{G-1}\frac{V}{W_s}\gamma_c(r - r_w) \times 100\% = \underline{3.24}\ (R - R_w)\ ;\ N' = \%\ \text{FINER NO. 200} \times N = \underline{.159}\ N \binom{\text{FOR COMBINED}}{\text{ANALYSIS ONLY}}$$

$$D\ \text{IN mm} = \sqrt{\frac{18\mu}{\gamma_s - \gamma_w}}\ \sqrt{\frac{Z_r}{t}} = \underline{.0129}\ \sqrt{\frac{Z_r\ \text{IN cm}}{t\ \text{IN min.}}}$$

DATE	TIME	ELAPSED TIME IN min.	R = 1000 (r − 1)	R_w = 1000 (r_w − 1)	TEMPERATURE IN °C	R − R_w	N IN %	Z_r IN cm	$\sqrt{\frac{Z_r\ \text{IN cm}}{t\ \text{IN min.}}}$	D IN mm	N'
10/14/49	14:00	0									
		¼	26.7	−0.5	22.6	27.2	88.1	11.1	6.65	.0860	14.0
		½	25.1			25.6	82.9	11.7	4.83	.0623	13.2
		1	21.0			21.5	69.7	13.3	3.64	.0469	11.1
		2	16.7			17.2	55.7	15.1	2.75	.0355	8.8
		2	16.5			17.0	55.1	13.9	2.64	.0340	8.8
	14:05	5	11.3			11.8	38.2	15.9	1.78	.0230	6.1
	14:10	10	8.4			8.9	28.8	17.1	1.31	.0169	4.6
	14:20	20	6.0			6.5	21.1	18.1	0.951	.0123	3.3
	14:40	40	4.6			5.1	16.5	18.6	0.682	.00880	2.6
	15:02	62	4.0	−0.5	22.6	4.5	14.6	18.8	0.550	.00710	2.3
	15:55	115	3.0			3.5	11.3	19.3	0.410	.00530	1.8
10/15/49	8:09	1089	1.5	−0.5	22.5	2.0	6.5	19.9	0.135	.00174	1.0
10/15/49	15:24	1524	1.2	−0.5	22.6	1.7	5.5	20.2	0.115	.00148	0.9

REMARKS:

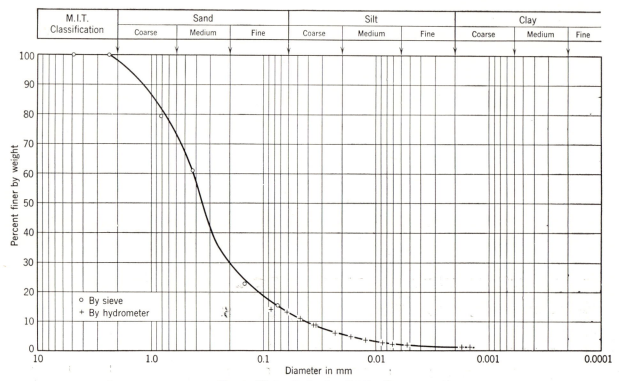

FIGURE IV-8. Grain size distribution.

CHAPTER

V

Compaction Test

Introduction

Many types of earth construction, such as dams, retaining walls, highways, and airports, require man-placed soil, or fill. To compact a soil, that is, to place it in a dense state, is desirable for three reasons: (a) to decrease future settlements, (b) to increase shear strength, and (c) to decrease permeability. Although the fundamentals of compaction are not completely understood, it is known that water plays an important part, especially in the finer-grained soils. As pointed out in Chapter III, soil particles adsorb a film of water when water is added to a dry soil. Upon the addition of more water, these films get thicker and permit soil particles to slide over each other more easily. This process is often called "lubrication." Since the thickness of a water film on a coarse particle is negligible in comparison with the particle diameter, lubrication effects are limited to the fine-grained soils.

Because of lubrication, the addition of a small amount of water to a dry soil aids the compaction process. Up to a certain point additional water replaces air from the soil voids, but, after a relatively high degree of saturation is reached, the water occupies space which could be filled with soil particles and the amount of entrapped air remains essentially constant. There is, therefore, an optimum amount of mixing water for a given soil and compaction process which will give a maximum weight of soil per volume.

The purpose of a laboratory compaction test is to determine the proper amount of mixing water to use when compacting the soil in the field and the resulting degree of denseness which can be expected from compaction at this optimum water content. To accomplish this purpose, a laboratory test which will give a degree of compaction comparable to that obtained by the field method used is necessary. In the early days of compaction, because construction equipment was small and gave relatively low densities,[1] a laboratory method that used a small amount of compacting energy was required. As construction equipment and procedures were developed which gave higher densities, it became necessary to increase the amount of compacting energy in the laboratory test.

In 1933, Proctor published (V–8) a series of four articles on soil compaction. In the second of this series, he described a laboratory compaction test which is now called the "standard Proctor" compaction test. Table V–1, which gives a comparison of some of the common laboratory tests, shows that the amount of compacting energy per volume used in the modified test is over four and one-half times that of the original Proctor test.

The critical question to be asked about any laboratory compaction test is, "How well does it represent field compaction?" Figure V–1 shows a comparison of field compaction against laboratory compaction for a silty clay (V–11). For this soil, the laboratory tests indicate optimum mixing water contents which are lower than the actual field optimum; this trend has been observed with most soils. Figure V–1 also illustrates the difficulty of choosing the proper laboratory test to use for a given soil and field compaction process.

Laboratory compaction tests are either dynamic or static; dynamic tests are used much more than static

---

[1] The word "density" is commonly used in compaction literature for a unit weight. The term "dry density" is defined on page 48.

43

ones. The results of the two types of tests are compared later in this chapter. None of the commonly used laboratory tests lend themselves well to the study of the compaction characteristics of clean sands or gravels. Although the regular laboratory methods are often used for these soils, their densities are not as significant functions of water content as the densities of the fine-grained soils, because of the negligible effects of lubrication on the coarse-grained soils. There

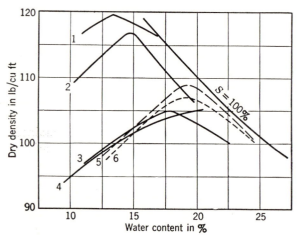

FIGURE V–1. Comparison of field and laboratory compaction.
(From reference V–11.)

(1) Laboratory static compaction, 2000 psi
(2) Modified AASHO
(3) Standard AASHO
(4) Laboratory static compaction, 200 psi
(5) Field compaction, rubber-tired load, 6 coverages
(6) Field compaction, sheep's-foot roller, 6 passes

Note: Static compaction from top and bottom of soil sample.

are laboratory compaction tests [2] used on coarse-grained soils which employ vibrations, but none, however, has gained wide acceptance at the present time.

With a knowledge of the moisture-density relationship as determined by a laboratory test, better control of the field compaction of the fill is possible because the optimum water content and the density which should be obtained using this water content are known. The engineer can conduct control tests, which are field determinations of water content and density, to insure proper compaction. The usual specification for field compaction is for the attainment of a certain percentage of the optimum density by a certain test; for example, a common one calls for "95% of the Modified AASHO optimum density."

[2] The U. S. Engineer Corps has suggested (V–16) that cohesionless soils be compacted in a saturated state with the modified AASHO compactive effort (see Table V–1) to obtain a density for field control.

Apparatus and Supplies

Special
  1. Compaction device
     (*a*) Mold 4.6 in. high, 4 in. diameter, $\frac{1}{30}$ cu ft volume.[3]
     (*b*) Removable mold collar 2.5 in. high, 4 in. diameter.
     (*c*) Hammer 2 in. diameter face, 5.5 or 10 lb weight (see Table V–1), and means for controlling its drop.

General
  1. Moisture sprayer
  2. No. 4 sieve
  3. Rubber-tipped pestle
  4. Scoop
  5. Straight edge and knife
  6. Large mixing pan
  7. Balances (0.01 lb sensitivity and 0.01 g sensitivity)
  8. Drying oven
  9. Desiccator
  10. Drying cans

Figure V–2 shows a manually operated compaction device of the controlled-fall type. As shown, it has the 10-lb hammer installed for the modified test and the replaceable 5.5-lb hammer for the standard test on the right (see Table V–1). The device is operated by raising the hammer with a pull on the cord and then permitting it to fall freely. Each of the two fins mounted on the vertical rod can be rotated into position to stop the hammer's upward movement and thus control its length of drop. Also shown in Fig. V–2 are a moisture sprayer, scoop, mixing pan, pestle, and knife.

A miniature compaction device has been made in which a mold of $1\frac{5}{16}$ in. inside diameter and 2.816 in. long is employed (V–18). The tamper has a prestressed spring inside the handle which controls the magnitude of the applied force. Laboratory compaction by this device can give results which more closely check field compaction than the commonly used apparatus; it has the additional advantages of requiring smaller test specimens [4] and of requiring less time for running a test. The miniature device appears promising and, after more research on the proper amount

[3] Molds of different dimensions are also used. See Table V–1 and the table footnotes. In his original publication, Proctor called for a mold "about 4 in. in diameter and 5 in. deep" (V–8).

[4] The use of small samples is convenient if fresh samples are used for each determination of water content and density. See page 46.

of compactive energy for different soils and different processes of field compaction, may become standard equipment in soil laboratories.

Another type of apparatus has been developed (V–5) which gives a kneading action in compacting the soil. This device can apply either static or dynamic compaction; and, like the device described above, it gives results which more closely agree

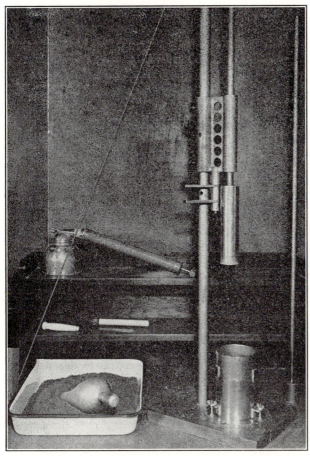

FIGURE V–2. Manually operated compaction device.

with field compaction than the commonly used ones do. It has the convenience of being operated automatically.

Recommended Procedure [5]

Since the commonly used dynamic compaction tests are alike except for the size of mold and compactive energy employed, only the standard Proctor or standard AASHO (American Association of State Highway Officials) test is presented below. This procedure can be easily altered to any of the other ones by changing the mold and/or energy in accordance with Table V–1.

---

[5] A student doing this test for the first time should be able to run a complete test in 2 to 3 hours and do the computations in about 1 hour.

1. Weigh the empty mold (with the base but without the collar) to 0.01 lb.

2. Obtain a 6-lb representative specimen of the soil sample [6] which is to be tested. Break all soil lumps in a mortar with a rubber-covered pestle and sieve the soil through a No. 4 sieve.

3. With the soil passing the No. 4 sieve, form a 2- to 3-in. layer in the mold.

4. Gently press the soil to smooth its surface and then compact it with twenty-five evenly distributed blows of the hammer, using a one-foot free drop. Between each drop of the hammer either the mold or the hammer should be rotated slightly to insure a uniform distribution of blows.

5. Repeat the procedure with a second and third layer, adjusting the drop of the hammer to one foot above the compacted soil layer. After compaction of the third layer, the surface of the soil should be slightly above the top rim of the mold.

6. Remove the collar and trim off the soil even with the top of the mold. In removing the collar, rotate it to break the bond between it and the soil before lifting it off the mold; this prevents removing some of the compacted soil when the collar is taken off. The trimming should consist of many small scraping operations with the straight edge, beginning at the central axis and working toward the edge of the mold.

7. After the soil has been made even with the top of the mold and all loose soil cleaned from the outside, weigh the cylinder and sample to 0.01 lb.

8. Remove [7] the soil from the cylinder and obtain a representative sample of approximately 100 g for a water content determination. The water content sample should be made up with specimens from the top, middle, and bottom of the compacted soil.

9. Break up by hand the soil removed [8] from the cylinder, remix with the original sample, and raise its water content approximately 3% [9] by adding water to the sample with the sprayer. Take care to distribute the water evenly and to mix the soil thoroughly. By weighing the sprayer before and after spraying, you

---

[6] The soil should be nearly air-dry for ease of sieving. However, on page 46 it is pointed out that higher densities may be obtained if the mixing water has been permitted to soak into the soil for some time.

[7] For removal, much time can be saved by extruding the soil from the mold by means of a jack or lever system. The sample extruder shown in Fig. I–2 can easily be adapted to fit the compaction mold, and thus used to push the compacted soil out of the mold.

[8] See page 46 for a discussion of the effect of reusing the compacted soil.

[9] For certain soils it may prove advisable to increase or decrease this approximate value of 3%.

can estimate the amount of water added. Knowledge of the water added helps you to control the moisture content.

10. Keep repeating the compaction process, each time raising the water content approximately 3%, until five or six runs have been made and the soil becomes very wet and sticky.

Discussion of Procedure

In Table V–1 a summary of commonly used compaction tests is given.

The tests of Table V–1, using the 6-in. diameter molds, are modifications made by the U. S. Engineer

compaction. Although the penetration needle appears to be gradually losing favor, a test based on a similar principle became popular during World War II. This test determines a penetration resistance of a compacted specimen of the soil in question which is compared to a standard resistance for crushed stone. The determined resistance divided by the standard for the stone multiplied by 100 is called the California Bearing Ratio (CBR). The CBR is widely used in semi-empirical methods of flexible pavement design (V–17).

Normally only soil passing a No. 4 sieve is used in the laboratory test (see step 3) because of the limited size of the mold. By discarding the coarser portion

TABLE V–1

Test	Mold Size *	Weight of Hammer, lb	Number of Layers	Height of Hammer Drop, in.	Number of Blows per Layer	Compactive Energy per Volume, ft-lb/cu ft
Standard Proctor Also Standard AASHO	4.6 in. x 4 in. diam.	5½	3	12	25	12,400
†	5± in. x 6 in. diam.	5½	3	12	55	12,400
Modified Proctor Also Modified AASHO	4.6 in. x 4 in. diam.	10	5	18	25	56,300
†	5± in. x 6 in. diam.‡	10	5	18	55	56,000
15-Blow Proctor	4.6 in. x 4 in. diam.	5½	3	12	15	7,400
15-Blow Proctor	5± in. x 6 in. diam.‡	5½	3	12	35	7,800

Porter static: a relatively high rate of compression until 100 psi is reached; compressed at the rate of 0.1 in. per minute until 1000 psi is reached; compressed at the rate of 0.05 in. per minute until 2000 psi is reached; the 2000 psi is maintained for one minute.

* Molds of slightly different dimensions but of the same volumes as those listed in this table are sometimes used.

† This test uses the same compactive energy per volume as the one listed above it and is, therefore, often given the same name (V–17), even though the mold size is different.

‡ The original mold of the California Bearing Ratio (CBR) was 5± in. x 6 in. diam., but in order to obtain approximately equal energies per volume as obtained in the small mold, the mold is only filled to a depth of 4½ in. This 4½ in. was used in computing the last column of the above table (V–17).

Corps (V–17) in order to obtain specimens large enough to determine the California Bearing Ratio.[10] The U. S. Engineer Corps based their modifications on the theory that the principal characteristic of a dynamic compaction test is the compactive energy per volume of soil. We can see from the last column in Table V–1 that this energy value has been kept essentially constant for each test when the mold size was increased.

In Figs. V–3 and V–4 are plotted compaction curves obtained by using different tests on a silty clay. Both these figures show that using more compacting energy tends to move the curves upward and to the left, a trend that exists on most soils.

In the early days of the compaction test, the force required to push a penetration needle, called a "soil plasticity needle," into the soil was used to indicate the density obtained in both laboratory and field

of the soil, we generally tend to obtain lower optimum densities and higher optimum water contents. This fact can be seen from Fig. V–5, which shows compaction curves for a well-graded sandy silt, with various **upper limits** of grain size being used.

In the procedure detailed in this chapter the same sample of soil was used to obtain all the density-water content measurements. The water for each stage was added, and compaction followed immediately. Research has shown that closer agreement with field conditions can be obtained if the water is allowed to soak for some time (e.g., overnight), and a fresh sample of soil is used for each determination. For example, it was found (V–9) that densities as much as 8 lb cu ft smaller were obtained by using fresh samples.[11] Allowing the water to soak in before compaction gives higher densities for many soils, particu-

[10] This ratio is explained in a following paragraph.

[11] Since the samples for this research project were not soaked prior to compaction, part of the low density may be explained.

larly for those with porous particles. The policy of some soil laboratories is to use fresh samples which have been soaked for each determination. This is a

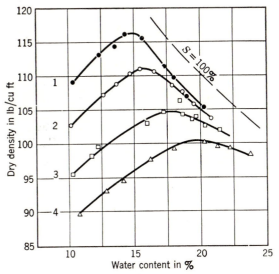

FIGURE V–3. Dynamic compaction curves for a silty clay. (From reference V–11.)

No.	Layers	Blows per Layer	Hammer Weight	Hammer Drop	
(1)	5	55	10 lb	18 in.	(mod. AASHO)
(2)	5	26	10	18	
(3)	5	12	10	18	(std. AASHO)
(4)	3	25	5½	12	

Note: 6 in. dia. mold used for all tests.

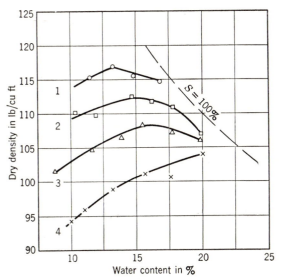

FIGURE V–4. Static compaction curves for a silty clay. (From reference V–11.)

  (1) 2000-psi static load
  (2) 1000-psi static load
  (3) 500-psi static load
  (4) 200-psi static load

Note: Compaction on top of soil sample.

commendable policy if enough soil is available. The objection to the large quantity of soil required for a test using fresh samples is overcome by the miniature compaction device described previously.

Whereas for some soils good agreement in shear strength has been obtained between laboratory and

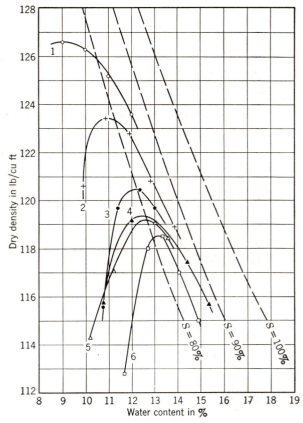

FIGURE V–5. Effect of maximum grain size on compaction. (From reference V–2.)

  (1) Passing ½ in.
  (2) Passing ⅜ in.
  (3) Passing No. 4 sieve
  (4) Passing No. 10 sieve
  (5) Passing No. 20 sieve
  (6) Passing No. 40 sieve

field compacted samples, when compared at equal water contents and densities, disagreement has been found for other soils. Even where agreement was noted, however, dissimilarly shaped stress-strain curves for the shear tests were obtained. More research is needed on this important question. There are some indications (V–10, V–17) that, if a soil is compacted at a given water content and then soaked until it reaches a higher water content, it is stronger than if it had been compacted at the higher water content. Compaction at the lower water content may impart some structural strength to the soil by affecting the water films on the particles.

Calculations

The dry density, γ_d (the weight of soil grains per unit of volume of soil mass), can be computed from [12]

$$\gamma_d = \frac{W}{V(1 + w)} \qquad \text{(V–1)}$$

in which W = total weight of moist compacted soil in cylinder,

V = volume of the mold,

w = water content of moist compacted soil.

For the original Proctor mold of $\frac{1}{30}$ cu ft volume,

$$\gamma_d = \frac{30W}{1 + w} \qquad \text{(V–1a)}$$

where W is in pounds and γ_d in pounds per cubic foot.

To plot the air void or degree of saturation curves, use

$$\gamma_d = \frac{G\gamma_w}{1 + (wG/S)} \qquad \text{(V–2)}$$

in which G = the specific gravity of the soil,

γ_w = unit weight of water,

S = degree of saturation.

The relation between dry density and void ratio, e, is

$$\gamma_d = \frac{G}{1 + e}\gamma_w \qquad \text{(V–3)}$$

Results

Method of Presentation. The results of a laboratory dynamic compaction test are usually presented in a plot of dry density versus water content (see Fig. V–6); the plot often includes curves of percentage of saturation or percentage of air voids. The air void curves represent the relationship of water content to dry density for given degrees of saturation. In other words, the broken line on the left in Fig. V–6 indicates the various dry densities which would be obtained at various water contents if the degree of saturation were kept at 80% or percentage of air voids at 20%. It is frequently useful to have scales of void ratio and porosity in addition to dry density. The results of compaction tests can be tabulated by giving the optimum density and water content for each test.

Typical Values. Based on the standard Proctor compaction test, a typical value of optimum dry density is 110 or 115 lb per cu ft obtained at a typical optimum water content in the neighborhood of 14%. The modified Proctor compaction test normally gives an optimum dry density which is 5% to 10% higher and an optimum water content which is a few percent lower than the corresponding values obtained by the standard Proctor compaction test (see Fig. V–3). The optimum dry densities vary over a wide range for different soils. An example of the possible extremes is given by the following two values: 69 lb per cu ft for a pumice and 140 lb per cu ft for a boulder clay. The better graded a soil is and the larger its largest particles are, the higher is its optimum dry density.

The density and water content values obtained from a compaction test are affected by test procedures which have not been standardized. Such details as the rigidity of support of the mold during compaction,[13] the amount of time the molding water has been in the soil before the compaction, and the number of times a soil has been compacted influence the test results as discussed previously.

Numerical Example

In the example on pages 50 and 51 are presented the results of a standard Proctor compaction test on a glacial till from Maine. It shows an optimum dry density of 119.4 lb per cu ft at a water content of 11.4%. Glacial tills make good cores for earth fill dams because of their low permeability, low compressibility, and high strength.

REFERENCES

1. American Society for Testing Materials, "Procedures for Testing Soils," July, 1950.
2. Chavez, Atanacio T., "An Investigation of Variables Affecting Compaction in Soils," Master of Science Thesis, Department of Civil Engineering, Massachusetts Institute of Technology, 1947.
3. "Compaction of Embankments," *Proceedings of the Highway Research Board*, 1938.
4. McRae, John L., "Effective Stresses and Pore-Water Pressures in Soils during Compaction," presented before the Annual Meeting, American Society of Civil Engineers, January, 1950.
5. Osterberg, J. O., "Testing Equipment and Research Activities of the Soil Mechanics Laboratory, Northwestern University," Northwestern University, Evanston, Ill., June, 1948.
6. Porter, O. J., "The Preparation of Subgrades," *Proceedings of the Highway Research Board*, Part II, p. 324, 1938.
7. Porter, O. J., "The Use of Heavy Equipment for Obtaining Maximum Compaction of Soils," *Technical Bulletin* 109, American Road Builders' Association, 1946.
8. Proctor, R. R., "Design and Construction of Rolled Earth Dams," *Engineering News-Record*, Aug. 31, Sept. 7, Sept. 21, and Sept. 28, 1933.

---

[12] The wet density is W/V; therefore, the dry density is equal to the wet density divided by $1 + w$.

[13] The ASTM (V–1) requires that during compaction the mold rest on a uniform, rigid foundation weighing 200 lb or on one of equivalent rigidity.

9. Sowers, G. F., and G. H. Nelson, "Effect of Re-using Soil on Compaction Curves," *Proceedings of the Highway Research Board,* 1949.

10. Tsien, Shou-I, "A Study of Structure on Compacted Soils," Doctor of Science Thesis, Department of Civil Engineering, Massachusetts Institute of Technology, 1946.

11. Turnbull, W. J., "Compaction and Strength Tests on Soils," presented before the Annual Meeting, American Society of Civil Engineers, January, 1950.

12. War Department, "Soil Testing Set No. 1 and Expedient Tests," *War Department Technical Bulletin* TB5–253–1, June, 1945.

13. Waterways Experiment Station, Soil Compaction Investigation, *Report* No. 1, "Compaction Studies of Clayey Sands," *Technical Memorandum* No. 3–271, Vicksburg, Miss., April, 1949.

14. Waterways Experiment Station, Soil Compaction Investigation, *Report* No. 2, "Compaction Studies on Silty Clay," *Technical Memorandum* No. 3–271, Vicksburg, Miss., July, 1949.

15. Waterways Experiment Station, Soil Compaction Investigation, *Report* No. 3, "Compaction Studies on Sand Subgrade," *Technical Memorandum* No. 3–271, Vicksburg, Miss., October, 1949.

16. Waterways Experiment Station, Soil Compaction Investigation, *Report* No. 5, "Miscellaneous Laboratory Tests," *Technical Memorandum* No. 3–271, Vicksburg, Miss., June, 1950.

17. Waterways Experiment Station, "The California Bearing Ratio Test as Applied to the Design of Flexible Pavements for Airports," July, 1945.

18. Wilson, Stanley D., "Comparative Investigation of a Miniature Compaction Test with Field Compaction," presented before the Annual Meeting, American Society of Civil Engineers, January, 1950.

SOIL MECHANICS LABORATORY

COMPACTION TEST

SOIL SAMPLE _Sandy Clay Silt: grayish brown; well graded, subrounded particles; mostly quartz & feldspar, some mica; some fine roots; glacial till (boulder clay)_

LOCATION _Union Falls, Maine_
BORING NO. _—_ SAMPLE DEPTH _1 to 2 ft._
SAMPLE NO. _Hollis 9_
SPECIFIC GRAVITY, G_s, _2.69_

TEST NO. _CO-3_
DATE _July 23, 1950_
TESTED BY _WCS_
TYPE TEST _Standard Proctor_
MOLD: VOLUME _1/30 cu. ft._
             WEIGHT _4.35 lb._

DENSITY

DETERMINATION NO.	1	2	3	4	5	6	7	8
WT. MOLD + COMPACTED SOIL IN lbs.	8.24	8.37	8.52	8.77	8.82	8.74	8.67	
WT. MOLD IN lbs.	4.35	4.35	4.35	4.35	4.35	4.35	4.35	
WT. COMPACTED SOIL IN lbs.	3.89	4.02	4.17	4.42	4.47	4.39	4.32	
WET DENSITY IN lbs./cf.	116.5	120.5	124.9	132.4	133.9	131.5	129.4	
DRY DENSITY, γ_d, IN lbs./cf.	110.5	112.5	114.8	119.4	117.5	113.4	109.9	
VOID RATIO, e	.520	.492	.462	.405	.428	.479	.527	
POROSITY, n	.342	.329	.316	.288	.300	.324	.345	

WATER CONTENT

DETERMINATION NO	1	2	3	4	5	6	7	8
CONTAINER NO	H-8	H-6	H-7	H-10	H-16	D-21	D-7	
WT. CONTAINER + WET SOIL IN g	212.4	159.3	133.7	134.4	124.9	114.6	140.2	
WT. CONTAINER + DRY SOIL IN g	204.1	152.0	127.1	126.4	115.5	104.4	125.2	
WT. WATER, W_w, IN g	8.3	7.3	6.6	8.0	9.4	10.2	15.0	
WT. CONTAINER IN g	55.1	47.6	52.7	53.4	47.9	40.7	41.4	
WT. DRY SOIL, W_s, IN g	149.0	104.4	74.4	73.0	67.6	63.7	83.8	
WATER CONTENT w, IN %	5.6	7.0	8.9	11.0	13.9	16.0	17.9	

PERCENT SATURATION

	w	γ_d			w	γ_d
	13 %	124.5 lb/cf			13 %	116.8 lb/cf
S = 100%	15	119.7		S = 80%	15	111.6
	17	115.3			17	106.9

REMARKS: _About 10% of soil retained on #4 sieve and not used in test._

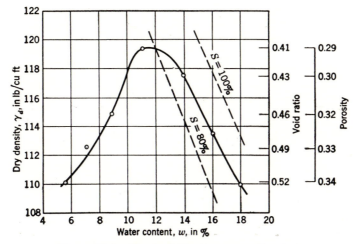

FIGURE V-6. Standard Proctor compaction test.

VI

Permeability Test

Introduction

A hundred years ago, Darcy showed experimentally that the rate of water q flowing through soil of cross-sectional area A was proportional to the imposed gradient i or

$$\frac{q}{A} \sim i \qquad q = kiA$$

The coefficient of proportionality k has been called "Darcy's coefficient of permeability" or "coefficient of permeability" or "permeability." [1] Thus permeability is a soil property which indicates the ease with which water [2] will flow through the soil.

Permeability enters all problems involving flow of water through soils, such as seepage under dams, the squeezing out of water from a soil by the application of a load, and drainage of subgrades, dams, and backfills. As will be discussed in later chapters, the effective strength of a soil is often indirectly controlled by its permeability.

Permeability depends on a number of factors. The main ones are:

1. *The size of the soil grains.* As pointed out on page 30, permeability appears to be proportional to the square of an effective grain size. This proportionality is due to the fact that the pore size, which is the primary variable, is related to particle size.

2. *The properties of the pore fluid.* The only important variable of water is viscosity, which in turn is sensitive to changes in temperature. Equation VI–3 expresses the relationship between viscosity and permeability.

3. *The void ratio of the soil.* The major influence of void ratio on permeability is discussed later in this chapter.

4. *The shapes and arrangement of pores.* Although permeability depends on the shapes and arrangement of pores, this dependency is difficult to express mathematically.

5. *The degree of saturation.* An increase in the degree of saturation of a soil causes an increase in permeability. This effect is illustrated by Fig. VI–1.

For testing sands and silts, the normal procedure is first to determine, by laboratory tests on disturbed samples, the relationship of void ratio to permeability. After obtaining the in situ void ratio of the soil, we can predict the in situ permeability by using the void ratio–permeability curve determined in the laboratory. This procedure is the most feasible one because of the difficulty of obtaining undisturbed samples of cohesionless soils. It should be remembered, however, that many soils have widely different [3] permeabilities along the stratification and perpendicular to it, and, therefore, the results obtained on disturbed samples may be of little real significance. The permeability of an undisturbed sample of clay can be determined directly at several different void ratios while running a consolidation test, as described in Chapter IX.

At least four laboratory methods of measuring the permeability of a soil are available. The variable

---

[1] The three terms are used interchangeably, even though the use here of "coefficient" may be questioned. The coefficient is not dimensionless, but has the units of velocity.

[2] The soil engineer rarely deals with pore fluids other than water. However, the permeability of a soil can also be obtained for fluids such as oil.

[3] Very frequently the permeability along the stratification is five to fifty times as large as that across it.

head and constant head tests are presented in this chapter. The capillarity method is presented in Chapter VIII; the use of consolidation test data to compute permeability is discussed in Chapter IX. The variable head test is normally more convenient for cohesionless soils than the constant head test because of the simpler instrumentation. There are conditions, however, under which the constant head test is preferable: for example, for the tests on partially

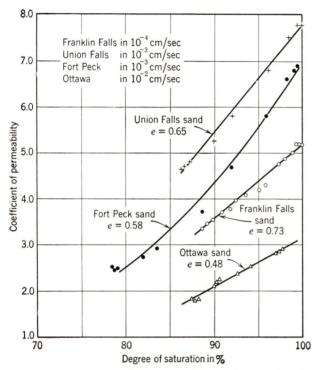

FIGURE VI-1. Permeability versus degree of saturation for various sands. (Data from reference VI-6.)

saturated soils (discussed in Chapter VIII) and for direct permeability determinations in conjuncton with consolidation tests (discussed in Chapter IX) on certain soils.

Apparatus and Supplies [4]
Variable Head Test
Special
1. Permeameter tube [5]
   (a) Two screens
   (b) Two rubber stoppers
   (c) Spring

---

[4] The apparatus for this test is described in more detail than for some of the other tests because it is more often constructed in the soils laboratory from stock materials.

[5] The desirable size of a permeameter depends on the soil to be tested. Permeameters in the neighborhood of 4 cm in diameter and 30 cm long have been found satisfactory for many soils. See page 58.

2. Standpipe
3. Deairing and saturating device
4. Support frame and clamps

General
1. Wooden hammer
2. Bell jar for constant head chamber
3. Supply of distilled, deaired water
4. Vacuum supply
5. Balance (0.1 g sensitivity)
6. Drying oven
7. Desiccator [6]
8. Scale
9. Thermometer (0.1° sensitivity)
10. Stop clock
11. Rubber tubing
12. Evaporating dish
13. Funnel
14. Pinch clamps

Figure VI-2 is a diagrammatic sketch of a variable head test setup which has proved satisfactory. In the laboratory, the parts can be permanently mounted to a panel or simply held to a support frame by clamps. The use of a transparent material, such as lucite, for the permeameter and water chamber is highly desirable, because it facilitates the measurement of the length of soil sample, L, and aids the detection of any air bubbles or movement of soil fines during the test. Likewise the water level in a transparent water chamber can be observed. The measuring of the soil length can be further facilitated by the cementing of graph paper strips, with units of length marked on them, to the outside of the permeameter. It is good policy to number each permeameter and standpipe, and mark on each its cross-sectional area. The bottom screen in the permeameter should be attached by some type of inside wedge and not screws, since screw holes are a possible source of leaks when the permeameter is evacuated.

The tubing should be either metal, high-pressure rubber (see Fig. VI-2), or some other material which can resist the applied vacuum. If low-pressure tubing is used between the standpipe and the permeameter, it will decrease in diameter as the hydrostatic pressure decreases because of a lowering of the water level in the standpipe. To prevent errors from such volume changes, the amount of tubing in this connection should be kept to a minimum. Water traps in the line preceding the manometers are desirable to prevent water from flowing into the manometers during the saturating process.

---

[6] A desiccator may not be needed. See page 10.

The choice of standpipe size should be made with regard to the soil to be tested. For a coarse sand, a standpipe whose diameter is approximately equal to that of the permeameter is usually satisfactory. On

listed for the variable head test. The additional items depend on the type of setup used.

In Fig. VI–3 are shown diagrammatically two test setups for running the constant head test. Although

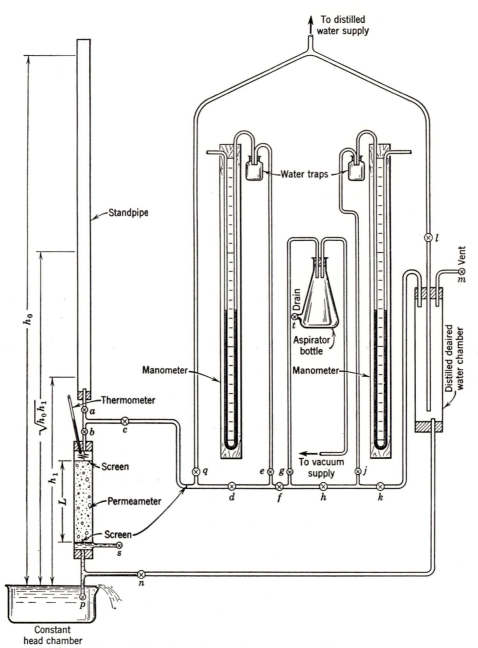

FIGURE VI–2. Setup for variable head permeability test.

the other hand, fine silts may necessitate a standpipe whose diameter is one-tenth or less of the permeameter diameter.

Constant Head Test. There are several items needed for the constant head test in addition to those

the one on the left is simpler, it should be used only for soils of high permeability. This limitation is due to the fact that, if the soil is relatively impermeable, the rate of flow is low, and thus the loss of water by evaporation can become an important consideration.

The balloons (Fig. VI–3b) furnish a convenient means of preventing evaporation. If the air inside them is allowed to become saturated with water vapor prior to testing, no evaporation will occur during the test (unless the atmospheric pressure or temperature changes). The balloons should be kept very loose so that the pressure in them will be essentially atmospheric.

If the diameter of the water supply bottle (Fig. VI–3b) is large relative to the diameter of the permeameter, the value of h can usually be considered

be applied to the water to obtain the additional head sometimes needed for testing impermeable soils.

Recommended Procedure [7]

The detailed procedures described below are for soils which are cohesionless; permeability determinations on fine-grained soils are discussed in Chapter IX.

Variable Head Test

1. Measure the inside diameter of the standpipe and permeameter.

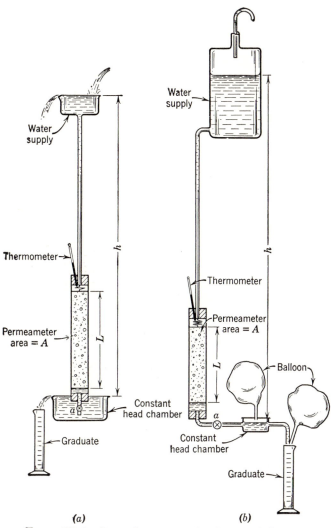

(a)   (b)
FIGURE VI–3. Setup for constant head permeability test.

constant for a test. The water level in the bottle should be recorded at the start and completion of the test to check the degree of validity of this assumption. The use of a bottle for the water supply has two advantages; it is a convenient means of storing water between tests, and it easily permits pressure to

2. Obtain to 0.1 g the weight of the empty permeameter plus screens, stoppers, and spring.

[7] A student doing this test for the first time should be able to test a cohesionless soil at three or four void ratios in 2 to 3 hours and do the computations in about an hour. He probably will need supervision for the first part of the test.

3. Load the permeameter with dry soil [8] to a loose, uniform density by pouring the soil in.[9]

4. Place the top screen, spring, and two stoppers in the tube. The spring should be compressed so that it will apply a pressure to the soil and help keep it in place when it is saturated.

5. Weigh the filled permeameter; the difference between the two weights is the amount of soil used.

6. Place the filled permeameter in position for testing as shown in Fig. VI–2.

7. Evacuate the sample to an absolute pressure of only a few centimeters of Hg by the following method:

(*a*) Close all valves shown in Fig. VI–2.

(*b*) Open valves *g*, *h*, *j*, *k*, *f*, *e*, *d*, *c*, and *b*.

8. After waiting some 10 to 15 minutes for the removal of air, saturate the soil by the following method:

(*a*) Close valves *f*, *g*, and *h*.

(*b*) Open valve *n*. The water will enter the soil because of the capillary attraction aided by the difference in elevation between water chamber and permeameter. If more head difference is needed, it can be obtained by slightly opening vent *m*. The difference in readings of the two manometers will indicate the additional pressure head that is thus obtained.

(*c*) Allow the water to saturate the sample and rise up to valve *b*, then close *n*.

(*d*) Release the vacuum on the sample by first closing *k* and *d*, and then slowly opening *q* and *m*.

(*e*) Any air bubble in the permeameter above the soil should be removed by slightly opening the upper stopper while applying water through *q*, with *d* closed. Any bubble in the bottom should be removed through *s*, while applying water through *n* with *m* open.

9. Measure the length of sample *L* and locate and measure the heads h_0 and h_1. The top limit of h_0 is selected at the upper end of the standpipe; h_1 a few centimeters above the lower end of the standpipe; the head $\sqrt{h_0 h_1}$ should be marked on the standpipe.

10. With valves *n* and *d* closed, fill the standpipe with distilled, deaired water to an elevation which is a few centimeters above h_0 by opening valves *q*, *c*, and *a*. Close valve *c*; leave *a* open.

11. Check to see that there is no air in the line between the standpipe and permeameter up to valve *c*, as

well as in the line from the permeameter into the constand head chamber.

12. Begin the test by opening valve *p*; start the timer as the water level falls to h_0 and record the elapsed times when the water level reaches $\sqrt{h_0 h_1}$ and h_1. Stop the flow after the level passes h_1 by closing *p*.

13. Obtain temperature readings at the head water end of the sample and in the constant head chamber.

14. Compare the elapsed time required for fall from h_0 to $\sqrt{h_0 h_1}$ with that for $\sqrt{h_0 h_1}$ to h_1.[10] If these times do not agree within 2% or 3%, refill and rerun.[11]

15. When a good run has been obtained, decrease the void ratio by tapping the side of the permeameter with the wooden hammer.

16. Remeasure the sample length and obtain time observations for the falling head in the standpipe as was done for the previous void ratio.

Constant Head Test

1. Place the soil in a measured permeameter, weigh, and saturate as in the variable head test (steps 1–8).

2. Measure the value of the head, *h*, and specimen length, *L*.

3. Start flow by opening valve *a* (see Fig. VI–3).

4. After allowing a few minutes for equilibrium conditions to be reached, obtain graduate and time readings.

5. After a sufficient amount of water has collected in the graduate for a satisfactory measure of its volume, take graduate and time observations. Subtract the graduate and time readings obtained in step 4 from the respective values obtained in this step to give *Q* and *t* for Eq. VI–2.

6. Record the temperature of the water every few minutes.

7. Change the void ratio of the soil as was done in the variable head test, and take another series of graduate and time readings. Measure the specimen length at each void ratio.

Discussion of Procedure

Degree of Saturation. In the preceding procedure, an attempt was made to get the soil completely satu-

---

[8] See page 58 for a discussion of the maximum grain size which should be used.

[9] Pouring the soil into the permeameter tends to cause segregation. Segregation can be minimized by placing the soil with a small can tied to strings in such a way that it can be lowered into the permeameter and then emptied.

[10] Since

$$\frac{h_0}{\sqrt{h_0 h_1}} = \frac{\sqrt{h_0 h_1}}{h_1}$$

the elapsed times should be equal because the other terms in Eq. VI–1 are constant for any given run. A lack of agreement here could be due to leaks, incomplete saturation, movement of fines, foreign matter in water, or water not sufficiently deaired.

[11] Even though the times for the two decrements are in agreement, it is a good policy to make a check run (see Numerical Example).

rated because the permeability of an "almost saturated" soil may be considerably different from its saturated [12] value. Figure VI–1 illustrates this point. To obtain a high degree of saturation, use a vacuum approaching absolute zero. For example, Fig. VI–4 shows the relationship between the degree of vacuum for evacuating a certain fine sand and the resulting degree of saturation. In this case an applied vacuum of at least 27 or 28 in. of mercury was necessary to get a high degree of saturation.

The water used for saturating the soil should be almost completely deaired, because if there is much

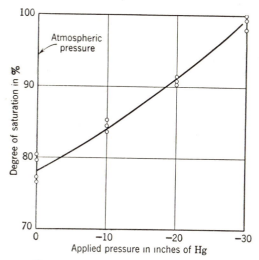

FIGURE VI–4. (From reference VI–4.)

air dissolved in the water, most of it will be brought out of solution by the high vacuum used for the saturating process of step 7 (see page 56). The deairing of the saturating water, however, presents no problems in the apparatus shown in Fig. VI–2. In fact, the procedure described in step 7 applies a vacuum to the water in the "distilled deaired water chamber" from which the saturating water is drawn. A vacuum can be kept on the water in this chamber when the apparatus is not in use.

Air dissolved in the water used for the actual permeability test causes no trouble in normal testing as long as it does not come out of solution to collect in the tubing or to collect in the soil, thus decreasing its degree of saturation. If water saturated with air were used, a rise in temperature or a decrease in pressure

---

[12] As discussed in Chapter VIII, natural soils do not necessarily exist in a saturated state. Careful control of the degree of saturation, however, is required in order to obtain test data which can be reproduced. Also, the permeability of a soil when saturated is a limiting value and, therefore, is of importance (see Chapter VIII). Unfortunately, there are permeability test procedures in use which do not control, or even measure, the degree of saturation.

in the water would have to be prevented as it flowed from its storage supply through the soil. This is because the solubility of air is proportional to the pressure of the air above the water for small pressures (Henry's law, VI–3) and decreases with temperature as shown by Fig. VI–5. The solubility of air in water may be altered by other changes in the water as it flows through the soil; for example, the dissolving of any soluble salts from the soil.

To prevent any air from coming out of solution, two procedures are recommended. First, keep the temperature of the water a few degrees warmer than the soil and tubing. If this is done, the water will cool as it flows, thus slightly increasing its capacity for dissolving air. This procedure is known as "maintaining a favorable temperature gradient." Second, use water which has less than its capacity of air dissolved in it; such water is commonly called "deaired" water.

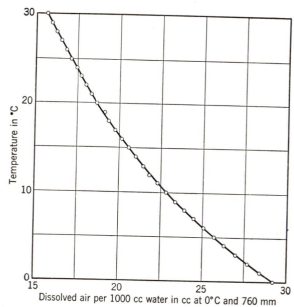

FIGURE VI–5. Solubility of air in water. *Note:* Air free of CO_2 and NH_3. (Data from *International Critical Tables*, Vol. III.)

Deaired Water. The air dissolved in water can be removed by increasing the temperature or decreasing the pressure. Boiling can reduce the dissolved air in water to about 0.75 ppm of oxygen or 1.5 cc of air.[13] Water which has been deaired is slow in regaining its air, as evidenced by Fig. VI–6, which is a plot [14] of

---

[13] One ppm of oxygen in air dissolved in water corresponds approximately to 2.0 cc of air at 760 mm pressure and 0° C per 1000 cc of water.

[14] This is a plot of data from a research project in the Hydraulics Laboratory at M.I.T. The data were obtained by the mercury-dropping electrode system; the readings were taken at

oxygen pick-up against elapsed time for a vessel of deaired water whose surface was exposed to the air. Figure VI–6 shows that at the end of 13 days the water was only 60% saturated. More elaborate methods for deairing and storing water are available (VI–2), but they are not thought necessary for normal permeability testing. Boiled distilled water is satisfactory for most permeability testing for some time after

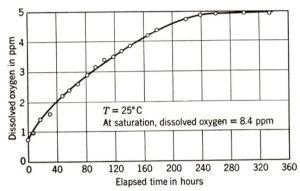

FIGURE VI–6. Pick-up of oxygen by water.

boiling. The water should not be agitated and should be covered to prevent the collection of foreign matter from the atmosphere. Water can easily be covered by stoppering the storage vessel and venting it with a tube whose end is pointing downward, as illustrated in Fig. VI–7. Figure VI–7 also shows the recommended manner to tap the water supply; the water at the bottom of the vessel tends to contain less dissolved air than that at the top.

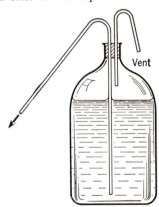

FIGURE VI–7. Storage of water for permeability tests.

Maximum Grain Size. To limit the maximum grain size of the soil tested to some reasonable fraction of the size of the permeameter is desirable. The use of large particles in a small permeameter increases

the chance of large voids forming where the particles touch the wall of the permeameter. Keeping the ratio of the permeameter diameter to the diameter of the largest soil particle greater than about 15 or 20 has been found satisfactory. This limits the soil tested in the 4-cm permeameter suggested on page 53 to that passing a No. 8 or No. 10 sieve. A larger permeameter should be used to test a coarser soil.

If the soil tested is too coarse, the flow will be turbulent rather than laminar. Laminar flow is assumed in Darcy's law, by which Eqs. VI–1 and VI–2 are derived. For the normal test setup, laminar flow exists only in soils finer than coarse sands. The error appears small, however, in using Darcy's law on soils whose particles are a little larger than coarse sand.

Gradient Increase by Gas Pressure. To increase the rate of flow in the constant head testing of soils of low permeability, a gas pressure can be applied to the surface of the water supply. (When a pressure is used, it is advisable to cover the surface of the water supply with a membrane of some sort to reduce the amount of gas going into solution.) The head lost is then h (Fig. VI–3) plus the applied pressure changed to units of water head. Pressure is often employed for permeability determinations on consolidation specimens (Chapter IX).

Calculations

Variable Head Test

The coefficient of permeability k can be computed from

$$k = 2.3 \frac{aL}{A(t_1 - t_0)} \log_{10} \frac{h_0}{h_1} \qquad (\text{VI–1})$$

in which a = cross-sectional area of the standpipe,
L = length of soil sample in permeameter,
A = cross-sectional area of the permeameter,
t_0 = time [15] when water in standpipe is at h_0,
t_1 = time when water in standpipe is at h_1,
h_0, h_1 = the heads between which the permeability is determined (see Fig. VI–2).

Constant Head Test

The coefficient of permeability k can be computed from

$$k = \frac{QL}{thA} \qquad (\text{VI–2})$$

in which Q = total quantity of water which flowed through in elapsed time, t,
h = total head lost (see Fig. VI–3).

---

a point ½ in. below the air-water interface in a vessel 5½ in. in diameter and 48 in. deep. The rate of air pick-up is related to the ratio of exposed surface area over volume of the water.

[15] If the time is started at zero when the water in the standpipe is at h_0, then t_0 is equal to zero.

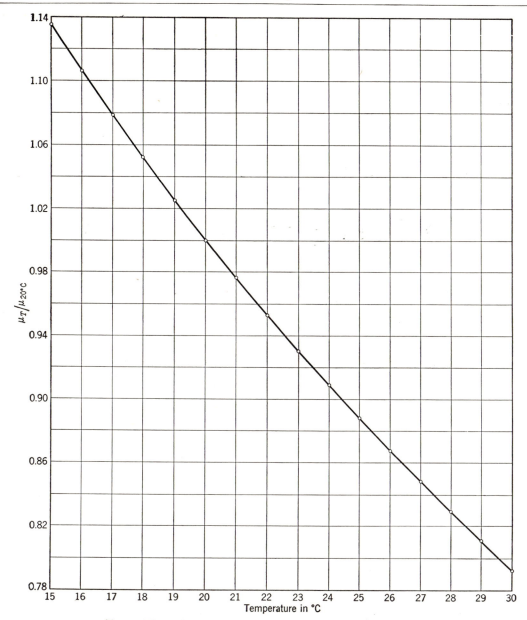

FIGURE VI–8. (Data from *International Critical Tables*, Vol. V.)

The permeability at temperature T, k_T, can be reduced to that at 20° C, $k_{20°C}$, by using

$$k_{20°C} = k_T \, \frac{\mu_T}{\mu_{20°C}} \qquad (VI-3)$$

in which $k_{20°C}$ = permeability at temperature 20° C,
        k_T = permeability at temperature T,
        μ_T = viscosity of water at temperature T (see Table A–3, p. 148),
        $\mu_{20°C}$ = viscosity of water at temperature 20° C (see Table A–3, p. 148).

A plot of $\mu_T/\mu_{20°C}$ against temperature is given in Fig. VI–8.

Results

Method of Presentation. The results of a permeability test are usually presented in the form of a plot of some function of void ratio, e, against some function of permeability, $k_{20°C}$. Often two plots are made: k vs. $e^3/(1 + e)$, $e^2/(1 + e)$, and e^2 on one sheet and e vs. log k. The best relationship of the above four is then used to present the results of the test (see discussion below).

Typical Values. The permeabilities of several soils are given in Fig. VI–1. A better indication of typical permeabilities can be obtained from the classification of soils based on their permeabilities which is given below (VI–5).

Degree of Permeability	k in Centimeters per Second
High	Over 10^{-1}
Medium	10^{-1} to 10^{-3}
Low	10^{-3} to 10^{-5}
Very low	10^{-5} to 10^{-7}
Practically impermeable	Less than 10^{-7}

A permeability of 1 μ per second (10^{-4} cm per second) is frequently used as the borderline between pervious and impervious soils. Thus a soil with a permeability less than 1 μ per second might be considered for a dam core or impervious blanket, whereas one with a permeability greater than 1 μ per second might be considered for a dam shell or pervious backfill.

Discussion. Both theoretically and experimentally there is more justification for $e^3/(1+e)$ to be proportional to k than for either $e^2/(1+e)$ or e^2 in the case of cohesionless soils. Laboratory tests on all types of soils have shown that a plot of void ratio versus log of permeability is usually close to a straight line.

Numerical Example

In the example on pages 61 and 62 are presented the results of a variable head permeability test on a well-graded, coarse sand, which was used for the shell of an earth dam. The plots of data in Fig. VI–9 show that $e^3/(1+e)$ is almost proportional to k and that the e vs. log k curve is almost a straight line. According to the classification given under Typical Values, this soil would be called one of medium permeability.

REFERENCES

1. American Society for Testing Materials, "Procedures for Testing Soils," Philadelphia, Pa., July, 1950.
2. Bertram, G. E., "An Experimental Investigation of Protective Filters," *Harvard University Publication* No. 267, 1939–1940.
3. Millard, E. B., *Physical Chemistry for Colleges*, McGraw-Hill Book Co., New York and London, 1946.
4. Parfitt, H. R., and N. E. Pehrson, "Experimental Investigation of the Degree of Saturation in Sands," Master of Science Thesis, Department of Civil Engineering, Massachusetts Institute of Technology, 1948.
5. Terzaghi, K., and R. B. Peck, *Soil Mechanics in Engineering Practice*, John Wiley and Sons, New York, 1948.
6. Wallace, M. I., "Experimental Investigation of the Effect of Degree of Saturation on the Permeability of Sand," Master of Science Thesis, Department of Civil Engineering, Massachusetts Institute of Technology, 1948.

SOIL MECHANICS LABORATORY

PERMEABILITY TEST

TEST NO. _P-3_
DATE _July 25, 1950_
TESTED BY _WCS_

STANDPIPE
NO. _1_ DIAMETER, d _4.09 cm_
AREA, a _13.17 cm²_

SOIL SAMPLE _Sand: brownish gray; coarse, well graded, subrounded particles; mostly quartz and feldspar, some mica._

LOCATION _Union Falls, Maine_
BORING NO. _CF_ SAMPLE DEPTH _5_
SAMPLE NO. _CF-2-5_
SPECIFIC GRAVITY, G_s _2.69_

$h_o = 140.0\ cm$
$h_1 = 70.0\ cm$

SOIL SPECIMEN WEIGHT
WT. PERMEAMETER + DRY SOIL IN g _778.0_
WT. PERMEAMETER IN g _324.4_
WT. DRY SOIL USED, W_s, IN g _453.6_

PERMEAMETER
NO. _B_
DIAMETER, D _4.42 cm_
AREA, A _15.32 cm²_

DETERMINATION NO.	SAMPLE LENGTH, L, IN cm.	TEMPERATURE, T, IN °C	ELAPSED TIME IN sec. FOR FLOW FROM h_o TO $\sqrt{h_o h_1}$	ELAPSED TIME IN sec. FOR FLOW FROM h_o TO h_1	PERMEABILITY AT T°C, k_T, IN cm./sec.	VISCOSITY at T / VISCOSITY at 20°C	PERMEABILITY AT 20°C, $k_{20°C}$, IN microns/sec.	VOID RATIO, e	e^2	$\dfrac{e^2}{1+e}$	$\dfrac{e^3}{1+e}$
1a	18.0	28.5	47¼	95	0.112	0.820	918	0.639	0.408	0.249	0.159
b	18.0	28.5	47¾	96¼							
2a	17.3	28.5	61½	123½	0.0823	0.811	668	0.578	0.333	0.211	0.122
b	17.3	29.0	61½	125							
3a	16.5	29.2	86½	174	0.0559	0.805	449	0.502	0.252	0.168	0.084
b	16.5	29.3	87½	176							
4a	16.0	29.2	111	223	0.0427	0.807	345	0.458	0.210	0.144	0.066
b	16.0	29.2	111½	224							

REMARKS: _Plus No. 4 screen material removed._

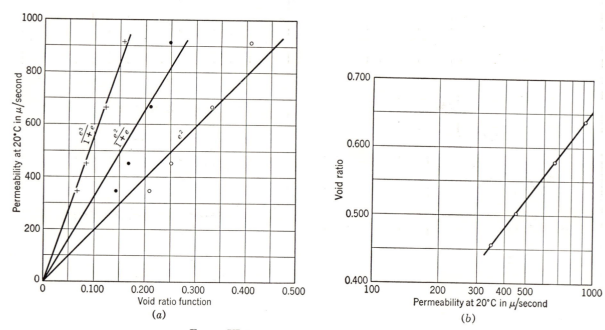

FIGURE VI-9. Variable head permeability test.

Introduction

There is much evidence that a liquid surface resists tensile forces because of the attraction between adjacent molecules in the surface. This attraction is measured by "surface tension," which is a constant property of any pure liquid at a given temperature. An example of this evidence is the fact that water will rise and remain above the line of atmospheric pressure, or phreatic line, in a very fine bore, or capillary, tube. This phenomenon is commonly referred to as "capillarity."

Capillarity enables a dry soil to draw water to elevations above the phreatic line; it also enables a draining soil mass to retain water above the phreatic line. The height of water column which a soil can thus support is called "capillary head" and is inversely proportional to the size of soil void at the air-water interface.[1] Since any soil has an almost infinite number of void sizes, it can have an almost infinite number of capillary heads. In other words, the height of water column which can be supported is dependent on the size of void that is effective. There is, therefore, no such thing as *the* capillary head for a soil; there are limiting values of capillary head which can best be explained by the setup in Fig. VII–1.

On the left of Fig. VII–1 is shown a tube of cohesionless[2] soil; on the right is shown a plot of degree of saturation against distance above the phreatic line. If the tube of soil were initially saturated and allowed to drain until a static condition was reached, the distribution of moisture could be represented by a line such as line *A*. If, on the other hand, the tube of dry

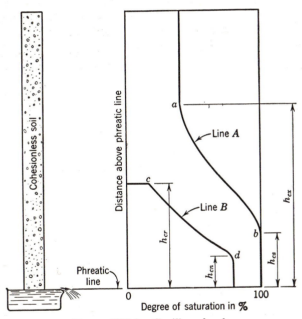

FIGURE VII–1. Capillary heads.

soil were placed in the container of water, line *B* would represent the distribution of moisture at equilibrium.[3] The lines *A* and *B* represent the two limiting conditions of capillary moisture distribution for the tube of soil shown.

---

[1] The height of rise, h_c, in a capillary tube (VII–1) is

$$h_c = \frac{2T_s}{R\gamma} \cos \alpha$$

where γ = unit weight of the liquid,
T_s = surface tension of the liquid,
α = contact angle made between the liquid and the tube,
R = radius of the tube.

[2] Most of the general concepts of capillarity presented here also hold for cohesive soils. The practical uses of capillary data in cohesive soils are limited.

[3] The times required for equilibrium to be reached depend greatly on the soil. Extremely long times are generally required to obtain line *B*. Actual times required for equilibrium for a fine sand as well as test data presented in the form of Fig. VII–1 can be found in reference VII–2.

It would seem logical that point *a* on the drainage curve (Fig. VII–1) is the highest elevation to which any continuous channel of water exists above the free water surface. This distance, therefore, is taken as the maximum capillary head, h_{cx}. Another critical point on the degree of saturation curve for a draining soil is the highest elevation at which complete saturation exists (point *b*, Fig. VII–1). The distance from the free water surface to this point is called the saturation capillary head, h_{cs}.

On the distribution curve for capillary rise, there are also two critical points. The distance from the free water surface to the highest elevation to which capillary water would rise (point *c*, Fig. VII–1) is called the capillary rise, h_{cr}. The distance from the free water surface to the highest elevation at which the maximum degree of saturation exists (point *d*, Fig. VII–1) is named the minimum capillary head, h_{cn}.

The four capillary heads described above are limits of the possible range of capillary heads a soil can have. Any capillary head associated with drainage would lie between h_{cx} and h_{cs}, and any associated with capillary rise would lie between h_{cr} and h_{cn}. Since the size of void at the air-water interface determines the capillary head, it is reasonable, in the case of a dropping water column, for a small void to develop a meniscus which can support the water in larger voids below its surface, while it could not possibly raise the water of a rising column past these larger voids. That h_{cx} should be greater than h_{cr} and h_{cs} than h_{cn}, therefore, is to be expected.

Between the two extremes, h_{cx} and h_{cn}, there exist many capillary heads. The effective capillary head for any soil problem involving capillarity would depend on the particular problem, but would lie within the range of limiting heads described above.

For comparing various soils and for certain drainage problems, the saturation capillary head, h_{cs}, is of much value. Since this head indicates the depth of soil below the water table which would undergo no loss of water after a lowering of this water table, it has direct application in design problems such as those involving the determination of lateral pressure on a wall retaining an earth fill. Not only is the saturation capillary head one of the more useful capillary heads, but it is also one of the easiest to measure. In the following pages, a procedure for measuring this head is presented.

Apparatus and Supplies

Special
1. Sample tube
   (a) Two screens
   (b) Two rubber stoppers
   (c) Spring
2. Head control chamber
3. Deairing and saturating device
4. Support frame and clamps

General
1. Tamper
2. Supply of distilled, deaired water
3. Vacuum supply
4. Balance (0.1 g sensitivity)
5. Drying oven
6. Desiccator [4]
7. Scale
8. Thermometer (0.1° C sensitivity)
9. Tubing
10. Evaporating dish
11. Funnel
12. Pinch clamps

The setup shown in Fig. VI–2 and described in Chapter VI can be used to saturate the soil. Figure VII–2 shows the setup after the sample has been saturated and the screen, spring, and stopper have been removed from the top of the sample container. The container, chamber, and scale (a meter stick or other type of rigid scale) can be conveniently held to a support frame or ring stand with clamps. Water (tap water is all right) is kept flowing continuously into the chamber through the inflow to maintain the water level in the chamber at the elevation of the top of the overflow drain.

The type of transparent tube described in Chapter VI is recommended for the sample container, although a tube of shorter length is satisfactory. The bottom screen of the container should be held in place by some sort of inside wedge because holes for screws are possible sources of leaks as the hydrostatic pressure within the container decreases during the test. A transparent, flexible, plastic tubing between the control chamber and the sample container is desirable, because it permits the detection of any air bubbles in the tubing.

Recommended Procedure [5]

This test consists of increasing the tension [6] in the pore water at the bottom of the soil until a bubble is drawn through the soil. The tension necessary to pull

[4] A desiccator may not be needed. See page 10.
[5] A student performing this test for the first time should be able to test a silty soil in 2 to 3 hours. He will probably need supervision for the first part of the test.
[6] Since atmospheric pressure is taken as zero pressure, any pressure less than atmospheric is called tension.

the first bubble through is the saturation capillary head. The steps are as follows:

1. Measure the inside diameter of the sample container.

2. Weigh the clean, dry, empty container to 0.1 g. Include the screens, spring, and top stopper.

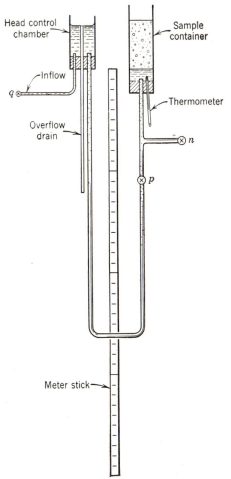

FIGURE VII–2. Setup for capillary head test.

3. Fill [7] the container with dry soil [8] to a height such that the top screen, spring, and stopper fit tightly when in place. The spring should be compressed so that the soil is kept in place when it is saturated.

4. Weigh the loaded container with top screen, spring, and stopper in place. This weight minus that obtained in step 2 is the weight of dry soil used.

5. Assemble the sample container as shown in Fig. VI–2 without any standpipe. Evacuate and saturate the sample as described in steps 7 and 8 of the variable head permeability test in Chapter VI. After saturation close valve n.

[7] See footnote 9, page 56.

[8] See page 58 for a discussion of the maximum allowable grain size of the soil sample.

6. Remove the stopper, spring, and screen from the top of the soil container and connect the head control chamber as shown in Fig. VII–2.

7. Measure the length of soil sample to an accuracy of 0.1 cm.

8. Make certain that there is no air in the line between head control chamber and the sample container, then open valves p and q.

9. Increase the water tension by lowering the head control chamber 2 cm [9] every 5 minutes. [10] Take a temperature observation every 15 or 20 minutes. The water tension applied to the pore water at the bottom of the soil is the difference of elevation between the bottom of the soil and the water level in the head control chamber. If zero on the scale is set at the elevation of the sample bottom, the scale reading at the water level in the control chamber is the applied water head.

10. Lower the control chamber until the first air bubble appears below the bottom screen.

Discussion of Procedure

The nature of this test is such that one detailed procedure and apparatus cannot be applicable for all soils. The bottom screen for the sample container should be selected with regard to the particular soil to be tested. If too fine a screen is used, the capillary head measured in the test will be that of the screen, or the soil and screen together, and not the soil alone. The coarsest screen that can retain the soil sample is recommended. A laboratory will often have on hand several containers, each fitted with a screen of different size for general testing. Likewise, the details of step 9 may be altered to fit a particular soil, as pointed out in footnotes 9 and 10.

The saturation capillary head for some fine-grained soils may be too large to be measured in the setup shown in Fig. VII–2. In testing such soils, mercury can be substituted for some of the water in the tube coming from the sample container and the control chamber not used. See Fig. VII–3. The water tension is then $b + 13.6a$, where b is the height of water

[9] For a soil possessing a large saturation capillary head, an excessive amount of time is required to run the test by increasing the head 2 cm every 5 minutes. The test time can be greatly shortened by starting the test at a tension estimated to be about one-half or three-quarters of the saturation capillary head. Also the head may be increased in increments greater than 2 cm in the early stages of the test.

[10] The time of 5 minutes is to allow the soil to drain to a state of static equilibrium. The time required for drainage naturally varies with the soil being tested. Five minutes is normally satisfactory for soils coarser than fine silts.

in the tube and a is the difference in lengths of the mercury columns. For such a setup a transparent tube is necessary for measuring b and a.

Calculations

The tension, expressed in water head, required to draw the first air bubble through the bottom screen is

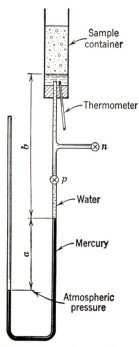

FIGURE VII–3. Setup for capillary head test.

the saturation capillary head. For the setup in Fig. VII–2, the tension is the difference in elevation between the water surface in the control chamber and the bottom of the soil; for the setup in Fig. VII–3, the tension is $b + 13.6a$.

Results

Method of Presentation. The results of the preceding test can be given by simply listing the saturation capillary head, the test temperature, and the void ratio of the soil specimen.

Typical Values. The test results on three typical soils are given below to indicate reasonable values of saturation capillary heads. These results are:

A coarse sand	10 cm
A uniform fine sand	60 cm
A glacial till	200 cm

If a dam which had a core made of the glacial till and a shell of the coarse sand were drained, the core would stay saturated by capillarity 200 cm or $6\frac{1}{2}$ ft above the phreatic line. The shell would be saturated for a distance of only 10 cm or 4 in. above the phreatic line.

Discussion. The value of saturation capillary head increases as the void ratio of the soil is decreased.

Although the temperature of the water in the saturation capillary head test is measured, usually no attempt is made to change the head to one at any particular temperature. Since the capillary head depends [11] directly on the surface tension of water, which in turn decreases with an increase of temperature, the capillary head decreases with an increase of temperature. However, the dependence of the capillary head on temperature is not thought important enough or understood well enough to justify attempts to correct for it.

Numerical Example

In the example on page 67 are presented the results of a capillary head test on the well-graded, coarse sand used in the permeability test example of Chapter VI. A saturation capillary head of 20 cm was obtained at a temperature of 21.1° C and a void ratio of 0.473.

REFERENCES

1. Freundlich, H., *Colloid and Capillary Chemistry,* Methuen and Company, Ltd., London, 1926.
2. Lambe, T. W., "Capillary Phenomena in Cohesionless Soil," Separate No. 4, American Society of Civil Engineers, January, 1950.

[11] The capillary head varies inversely with the unit weight of the water. The effect of temperature on the unit weight of water is small compared to the effect on the surface tension.

SOIL MECHANICS LABORATORY

CAPILLARY HEAD TEST

TEST NO. CH-1

DATE Jan. 10, 1950

TESTED BY DD

PERMEAMETER

NO. D ; DIAMETER, D 4.34 cm

AREA, A 14.8 cm²

SOIL SAMPLE Sand: brownish gray; coarse, well graded, subrounded particles; mostly quartz & feldspar, some mica.

LOCATION Union Falls, Maine

BORING NO. CG SAMPLE DEPTH 9

SAMPLE NO. CG-2-9

SPECIFIC GRAVITY, G_s, 2.69

SOIL SPECIMEN WEIGHT

WT. PERMEAMETER + DRY SOIL IN g 394.3

WT. PERMEAMETER IN g 167.1

WT. DRY SOIL, W_s, IN g 227.2

SOIL SPECIMEN VOLUME

SAMPLE LENGTH, L, IN cm 8.4

SAMPLE VOLUME, V, IN cc 124.2

SOLIDS VOLUME, V_s, IN cc 84.3

VOID RATIO, e .473

POROSITY, n 32.1%

TIME IN hr.-min.	ELEVATION BOTTOM OF SOIL IN cm	ELEVATION WATER LEVEL IN cm	TEMPERATURE, T, IN °C
3-50	0	0	21.0
3-55		-2	
4-00		-4	
4-05		-6	
4-10		-8	
4-15		-10	21.0
4-20		-12	
4-25		-14	
4-30		-16	
4-35		-18	
4-40	0	-20	21.1

h_{cs} = 20 cm

REMARKS:

TIME IN hr.-min.	ELEVATION BOTTOM OF SOIL IN cm	ELEVATION WATER LEVEL IN cm	TEMPERATURE, T, IN °C

CHAPTER

VIII

Capillarity-Permeability Test

In Chapter VII was presented the fact that capillarity enables a dry soil to draw water to elevations above the phreatic line, or line of atmospheric pressure, and also enables a draining soil to retain water above the phreatic line. The movement of water pulled into dry soil by capillarity is not smooth, but jerky, because the water travels more rapidly in the small pores, where the pull is stronger, than in the large pores. Because of this unsteady flow some of the pores are by-passed and encompassed, and thus remain filled with air. Figure VII–1 illustrates the fact that the maximum degree of saturation for capillary rise is less than 100%. If the capillary flow occurs in a horizontal tube of soil, thereby eliminating the effects of a varying elevation head, the degree of saturation back of the advancing wetted surface appears to be approximately the same as that for the height h_{cn} in the capillary rise test (see Fig. VII–1). Figure VIII–1 presents test data on a fine uniform sand to illustrate this point.[1] Because the degree of saturation in the region just behind the advancing wetted surface is lower than that ultimately attained (see Fig. VIII–1), the gradient is larger just behind the wetted surface than it is farther back.[2]

The degree of saturation in a cohesionless soil in nature below the phreatic line normally would be between complete saturation and the capillary degree of saturation (that obtained by capillary flow as illustrated in Fig. VIII–1). This degree of saturation can

[1] Tests on other soils indicate that a percentage of saturation in the range of 80% is typical for horizontal capillary flow in fine sands and silts (VIII–3).

[2] In reference VIII–2 measurements of pore water pressure are presented which show a much larger gradient in the zone just behind the advancing wetted surface.

be increased during flow by the water dissolving the entrapped air (VIII–3). As pointed out in Chapter VI, the water can also release dissolved air if its temperature is increased or pressure decreased. Thus seasonal changes may occur in the degree of satura-

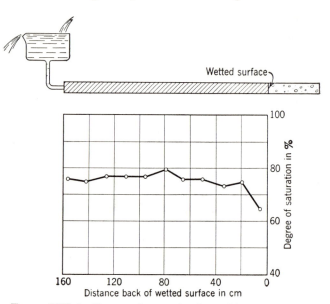

FIGURE VIII–1. Degree of saturation in capillary flow. (From reference VIII–2.)

tion of a soil in situ. Such changes, if they exist, are probably slight and would hardly reduce the degree of saturation to less than the capillary value.

Since the permeability of a soil is a function of the degree of saturation (see Fig. VI–1), a soil below the phreatic line should be expected to have an effective permeability ranging between that at complete saturation, k_{100}, and that at the capillary degree of saturation, k_s. The former can be approximated from the test described in Chapter VI, and the latter from the capillarity-permeability test described in this chapter. Again, it should be emphasized that stratification effects in a soil in nature may be much more important in influencing the effective permeability than the degree of saturation.

The capillarity-permeability test uses the setup shown in Fig. VIII–2. For the first stage of the test, h_{01} is applied at the left end of the soil sample and, for the second stage, h_{02} is applied; Eq. VIII–1, page 71, can be written for each stage. Simultaneous solution of these two equations gives the two unknowns, the effective capillary head, h_c', and the permeability, k_s. As mentioned above, a larger gradient exists just behind the wetted surface than farther back, where the degree of saturation is more nearly constant. Because Eq. VIII–1 is based on the assumption of a uniform gradient, this test does not supply a true capillary head but instead the water tension required to give the uniform gradient. Since this head depends on the details of the test setup (length of soil sample, magnitude of applied head, etc.) and has no real meaning, it cannot be compared to the capillary heads discussed in Chapter VII. After completion of the capillary test, the permeameter of soil can be mounted in the setup used in Chapter VI, and a check determination of the permeability can be made by either the variable or the constant head method.

Two soil properties obtained from the capillarity-permeability test are of value. These are the capillary degree of saturation, and the permeability at this degree of saturation. These two properties are used in seepage problems and in problems in which the degree of saturation is needed to compute soil weights.

Apparatus and Supplies

Special

1. Permeameter
   (*a*) Two screens
   (*b*) Two rubber stoppers
   (*c*) Spring
2. Standpipe
3. Support frame and clamps

General

1. Tamper
2. Bell jar for constant head chamber
3. Supply of distilled, deaired water
4. Balance (0.1 g sensitivity)
5. Drying oven
6. Desiccator [3]
7. Scale

[3] A desiccator may not be needed. See page 10.

8. Thermometer (0.1° C sensitivity)
9. Timer
10. Rubber tubing
11. Evaporating dish
12. Funnel
13. Pinch clamps

A sketch of the test setup is shown in Fig. VIII–2. The permeameter used for this test is identical with that used for the permeability test, except that it is longer.[4] For the horizontal capillary test, a permeameter of some transparent material, such as lucite, is essential in order that the position of the wetted surface can be determined. Strips of graph paper,

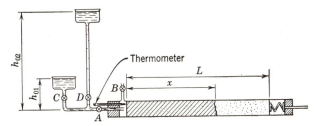

FIGURE VIII–2. Setup for horizontal capillary flow.

with units of length marked on them, are cemented in grooves cut into the outside of the tube to facilitate the observations of distance. It is desirable to place such strips every 120° or 90° around the circumference of the tube and to protect them with a coat of varnish.

The setup shown in Fig. VIII–2 can be simplified by using one large bottle instead of the two reservoirs. The applied head is then changed (step 9 of the Recommended Procedure) by moving this bottle to a higher elevation. If a bottle of large diameter is used for the water supply, the depth of water in the bottle is essentially unchanged by the small amount which flows into the soil. The validity of assuming a constant depth can be easily checked by measuring the depth of water in the bottle prior to and at the conclusion of testing. If there is an important difference between the two measurements, their average should be used.

Recommended Procedure [5]

1. Measure the inside diameter of the permeameter.
2. Weigh to 0.1 g the empty permeameter plus screens, stoppers, and spring.

[4] A permeameter about 4 cm in diameter and 35 cm long has been found satisfactory for many sands and silts.

[5] This test can be done better by a student group or at least two students. They should be able to do their first test in approximately 2 hours if they have previously run a permeability test. They probably will need supervision for the first part of the test.

3. Fill [6] the permeameter with dry soil [7] to a uniform, compact density (e.g., by tamping each spoonful of soil a given number of times).

4. Place the top screen, spring, and two stoppers in the tube. Compress the spring to minimize the chances for soil movement and push the stoppers in tightly to prevent their blowing out.

5. Weigh the filled permeameter; the difference between the weights of the permeameter when filled and when empty is the amount of soil used.

6. Measure the length of sample, L (see Fig. VIII–2); locate and measure the heads, h_{01} and h_{02}. Make h_{01} about 25 cm and h_{02} about 200 cm.[8] Make sure that there is no air in the tubing connecting the reservoirs to the permeameter.

7. Hold the permeameter vertical; start flow by opening valves C and A. As soon as the water reaches the soil, restore the permeameter to its horizontal position and start the timer.

8. Take readings of elapsed time at every full centimeter value of x. If necessary, roll the permeameter around a horizontal axis to keep the wetted surface approximately vertical.

9. When x is approximately equal to $L/2$, change from h_{01} to h_{02} by closing C and opening D, and continue to take readings until all the soil is wet. If at any time during the test, air collects between the stopper and screen at the entrance end, release it by slightly opening valve B.

10. Make sure that the space between the screen and stopper at both ends is completely filled with water, and place the permeameter in the support frame for a permeability test (see Fig. VI–2 or VI–3).

11. When in position, the entrance end (where valve A is) should be on top.[9] Measure the permeability by either the variable or constant head procedures given in Chapter VI. The sample is not saturated but tested at its existing degree of saturation (i.e., start with step 9 of variable head test or step 2 of the constant head test).

12. After completing the test, hold the tube horizontal [10] and remove both stoppers and the spring.

[6] See footnote 9 on page 56. Stratification can be reduced by scarifying the top of each layer before placing the next.

[7] See page 58 for a discussion of the maximum allowable grain size of the test specimen.

[8] The desirable values of h_{01} and h_{02} depend to some extent on the soil being tested. For many soils 25 and 200 cm have been found to work well. See further discussion of this point in the following paragraphs.

[9] This is desirable in order that the direction of flow for the two tests should be the same. Reversing the direction of flow may tend to move entrapped air bubbles, thus causing a change in the distribution of entrapped air.

[10] The permeameter of soil is held horizontal to minimize drainage of the pore water.

Carefully dry the spring, stoppers, the outside of the tube, and the inside of the tube which is not covered with soil. In other words, remove all water except that in the soil.

13. Weigh the wet soil, permeameter, screens, stoppers, and spring.

Discussion of Procedure

Because of the nature of the capillarity-permeability test, partially saturated soils must be tested; such testing requires a very highly developed laboratory technique to obtain correct results. The presence of air in soil or in the water lines of the apparatus is usually a source of trouble for even the best technicians. One example of the effects of air can be seen on page 73, where the slope of the t vs. x^2 curve does not change exactly at the time the value of h_0 was changed. The increase in hydrostatic pressure from the heightened applied head caused a compression of the entrapped air. The time lag resulted because water had to flow in to compensate for the air compression.

Because an increase in hydrostatic pressure compresses the entrapped air, the degree of saturation within a partially saturated soil is a function of hydrostatic pressure. For consistent results, therefore, applied heads should be kept within the range of those recommended in step 6. A larger value of h_{02} may change the degree of saturation enough to invalidate the assumption of constant degree of saturation in Eq. VIII–1.

The dependence of the degree of saturation on the applied pressure necessitates caution in the falling head test. The test should be started with the water in the standpipe falling past the upper mark.[11] If the initial point is taken when the water in the standpipe is in a static condition, that is, with the valve closed, then the entrapped air bubbles will increase in volume when the valve is opened because of the decrease in hydrostatic pressure. Such changes of air bubble size result in a time lag, similar to that shown in the plot on page 73 and described above. Figure VIII–3 is a plot of pore water pressure data for a permeability test on a partially saturated soil. A sketch of the test setup is given at the right of the figure; the pore water pressures were measured by means of piezometers, whose elevations are given adjacent to their respective plots. When the water had reached h_2 in the standpipe at an elapsed time of $36\frac{1}{2}$ minutes, the flow was stopped and the standpipe refilled to h_0. As Fig. VIII–3 shows, some 45 minutes were required

[11] The standpipe should be filled to its top in order to allow more time for pressure adjustment before the level reaches the upper limit of h_0.

for the change of entrapped air volume to occur; the degree of saturation of the soil increased from 80% to 84% during the 45 minutes.

For this soil, permeability values ranged from 3.91 to 4.42 μ per second, depending on the hydrostatic pressure existing in the pore water at the start of the test. Because of the difficulties of controlling the hydrostatic pressure in the pore water during the variable head permeability test, more consistent results can be obtained from the constant head test.

At the conclusion of the test the sample tube (step 13) should be weighed without delay because delay permits the air bubbles to expand and extrude water.

ments made during the test by using Eq. I–4, page 13. The permeability of the soil at the capillary degree of saturation, k_s, can be obtained from

$$\frac{\Delta(x)^2}{\Delta t} = \frac{2k_S}{Sn}(h_0 + h_c') \qquad \text{(VIII–1)}$$

in which $\dfrac{\Delta(x)^2}{\Delta t} = m =$ the slope of a plot of x^2 vs. t (see Numerical Example),

$n =$ porosity of the sample,
$S =$ capillary degree of saturation,
$h_c' =$ an effective capillary head,
$h_0 =$ the applied static head.

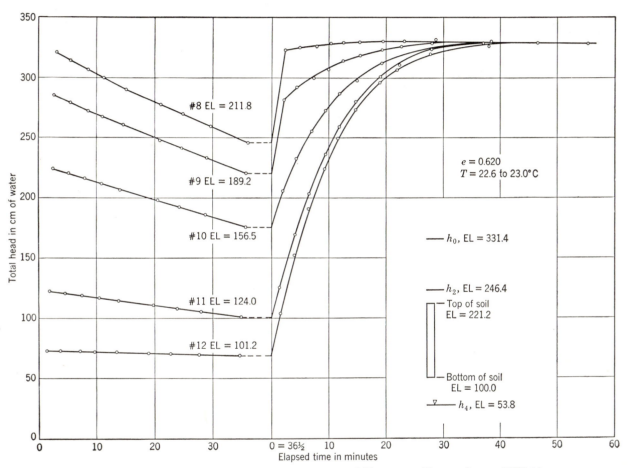

FIGURE VIII–3. Pore water pressures during a permeability test. (From reference VIII–1.)

Since the pore water pressure is lower during weighing (step 13) than during testing, there is a tendency to get a degree of saturation value lower than the effective one. On the other hand, incomplete drying of the permeameter prior to weighing tends to give a degree of saturation which is too large.

Calculations

The degree of saturation obtained by capillarity can be calculated from the weight and volume measure-

Two equations can be set up from the expression above by using both pairs of h_0 and m values. The simultaneous solution of the two equations results in the determination of the two unknowns, h_c' and k_S. Equation VI–3, page 59, can be used to reduce k_S to the permeability at 20° C, $k_{S\,20°C}$.

Results

Method of Presentation. The results of a capillarity-permeability test can be given by listing the capillary degree of saturation, permeability, and void ratio.

Typical Values. Because the effective capillary head, h_c', is not a fundamental soil property but a fictitious number, its magnitude has no meaning. If the soil permeameter is very long, values of h_c' which are in excess of atmospheric pressure can be obtained. The value of h_c' is always less than the true tension in the water just behind the wetted surface because of the lower degree of saturation existing at that point.

Tests on several sands and silts indicate that the capillary degree of saturation is usually in the range of 80%. As illustrated by Fig. VI–1, the permeability of a soil at 80% appears to run from one-third to one-half of that for the saturated soil.

Numerical Example [12]

A permeameter of approximately 35 cm is usually long enough for this test. A much longer one, however, was used for the example on page 73, in order to

[12] Since an example of a variable head permeability test was given in Chapter VI, that part of the test is omitted from the example.

show that essentially straight lines can be obtained for plots of x^2 against t for long samples.

The following results were obtained from the two parts of the test on a fine sand at a void ratio of 0.602 and degree of saturation of 74.5%.

	Horizontal Capillary	*Variable Head*
$k_{S\ 20°C}$	0.0254 cm/minute	0.0264 cm/minute

That agreement is considered good. The discrepancy is probably due to the difference in degree of saturation caused by the different hydrostatic pressures existing in the two tests.

REFERENCES

1. Lambe, T. W., "The Measurement of Pore Water Pressures in Cohesionless Soils," *Proceedings of the Second International Conference on Soil Mechanics,* Paper II a 9, Vol. VII, June, 1948.
2. Lambe, T. W., "Capillary Phenomena in Cohesionless Soil," published as a Separate No. 4, American Society of Civil Engineers, January, 1950.
3. Wallace, Milton I., "Experimental Investigation of the Effect of Degree of Saturation on the Permeability of Sand," Master of Science Thesis, Department of Civil and Sanitary Engineering, Massachusetts Institute of Technology, 1948.

SOIL MECHANICS LABORATORY

CAPILLARITY - PERMEABILITY TEST

SOIL SAMPLE _Sand: brown; fine,_
uniform, subrounded particles;
mostly quartz.

LOCATION __Union Falls, Maine__
BORING NO. _CE_ SAMPLE DEPTH _6_
SAMPLE NO. __CE-2-6__
SPECIFIC GRAVITY, G_s, __2.67__

HEADS

h_{01} _____35.0 cm_____
h_{02} _____181.7 cm_____

PERMEAMETER

NO. __C__ DIAMETER, D _4.38 cm_
AREA, A ____15.05 cm²____

SOIL SPECIMEN WEIGHT

WT. PERMEAMETER
+ DRY SOIL IN g ____9682____

WT. PERMEAMETER
IN g ____5112____

WT. DRY SOIL,
W_s, IN g ____4570____

WT. PERMEAMETER
+ WET SOIL IN g ____10449____

WT. WATER,
W_w, IN g ____767____

TEST NO. ____CP-2____

DATE ____Jan. 3, 1950____

TESTED BY ____DD____

SOIL SPECIMEN VOLUME

SAMPLE LENGTH, L, IN cm ___182.4___
SAMPLE VOLUME, V, IN cc ___2740___
SOLIDS VOLUME, V_s, IN cc ___1710___
VOID RATIO, e _____.602_____
POROSITY, n _____37.6%_____

DISTANCE TRAVELED, X, IN cm	x^2 IN cm²	ELAPSED TIME, t, IN min. - sec.	TEMPERATURE, T, IN °C
6	36	4-0	21.7
13	169	15-10	
15	225	20-45	
24	576	50-55	
28	784	69-35	21.8
30	900	79-45	
32	1024	90-20	
33.5	1122	99-00	
35	1225	106-15	
37	1369	122-20	
39	1521	135-20	
42	1764	155-20	
44	1936	169-20	22.0
45	2025	179-20	
47	2209	197-35	
48	2304	205-45	
53	2809	229-30	
55	3025	234-55	

DISTANCE TRAVELED, X, IN cm	x^2 IN cm²	ELAPSED TIME, t, IN min. - sec.	TEMPERATURE, T, IN °C
58.4	3411	244-15	
60	3600	249-00	
63	3969	257-20	
65	4225	264-35	22.3
68.3	4665	274-50	
70	4900	281-10	
73.5	5402	294-15	
77.5	6006	309-45	
80	6400	318-30	
83	6889	330-45	
85	7225	339-00	
88	7744	353-30	
90.2	8136	364-10	22.4

REMARKS:

Changed from h_{01}
to h_{02} at t = 215-15

$m_1 = 11.2 \frac{cm^2}{min}$

$m_2 = 39.3 \frac{cm^2}{min}$

S = 74.5 %

$h_c' = 23.7$ cm

$k_{s20°C} = .0254 \frac{cm}{min}$

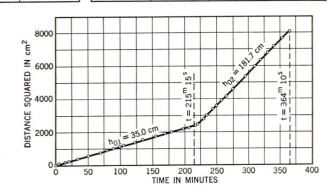

CHAPTER

IX

Consolidation Test

Introduction

When a saturated soil mass is subjected to a load increment, the load is usually carried initially by the water in the pores because the water is incompressible in comparison with the soil structure. The pressure which results in the water because of the load increment is named "hydrostatic excess pressure" because it is in excess of that pressure due to the weight of water. As the water drains from the soil pores, the load increment is shifted to the soil structure. The transference of load is accompanied by a change in the volume of soil equal to the volume of water drained. This process is known as "consolidation." [1]

We can be aided in understanding the consolidation process by the spring analogy shown in Fig. IX–1. The saturated soil element is represented by Fig. IX–1a, in which the spring corresponds to the soil structure and the water to the soil pore water. If a weight W is placed on the water and spring with the valve y closed (Fig. IX–1a), the weight is almost entirely carried by the water, since it is incompressible as compared to the spring. If valve y is opened and the water is allowed to escape, the load will eventually be carried entirely by the spring (Fig. IX–1c). The elapsed time required to transfer the load increment W from the water to the spring depends on how rapidly the water is permitted to escape through valve y.

The rate at which the volume change, or consolidation, occurs in a soil is directly related to the permeability of the soil because the permeability controls the speed at which the pore water can escape. The permeability of most sands is so high that the time required for consolidation after a load application can be considered negligible except for cases where a large

---

[1] Consolidation is often used in a less rigorous sense to define any volume decrease of a soil mass.

mass of sand is subjected to a rapid shear or shock. (This is discussed in Chapter X.) On the other hand, the low permeability of clay makes the rate of volume change after a load application a factor which must be considered. Laboratory consolidation studies, therefore, are almost entirely limited to soils of low permeability.

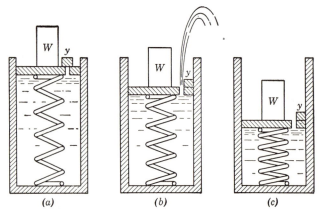

FIGURE IX–1. Spring analogy to consolidation.

Consolidation is an important fundamental phenomenon which must be understood by everyone who attempts to gain a knowledge of soil behavior in engineering problems. As will be pointed out in the following chapters, that portion of the strength of a soil created by friction depends directly on the pressure acting between the particles, that is, the intergranular pressure. In soil problems, this intergranular pressure is usually ascertained most conveniently by subtracting the water pressure from the combined, or intergranular plus water, pressure. In order to determine the shear strength of a soil, therefore, the stage of consolidation of the soil must be known. A study of the different types of shear tests in clay described in Chapter XIII emphasizes the importance of consolidation in shear.

As stated in Chapter VI, the consolidation test provides a convenient opportunity for a direct measurement of permeability by either the variable head or the constant head method. Because these methods are described at length in Chapter VI, only those details which must be altered will be discussed in this chapter. In addition, the permeability of the specimen can be computed from the regular consolidation test data (see Eq. IX–6, page 83).

The main purpose of consolidation tests, however, is to obtain soil data which are used in predicting the rate and the amount of settlement of structures founded on clay. Although some of the settlement of a structure on clay may be caused by shear strain, most of it is normally due to volume change,[2] particularly if the clay stratum is a thick one or one at some depth below the structure.

The two most important soil properties furnished by a consolidation test are (1) the compression index, C_c, which indicates the compressibility of the specimen, and (2) the coefficient of consolidation, c_v, which indicates the rate of compression under a load increment. The data from a laboratory consolidation test make it possible to plot a stress-volume strain curve, which often gives useful information about the pressure history of the soil. This stress-strain curve and the soil properties mentioned above are discussed and illustrated later in this chapter.

In order to predict the settlements of structures in the field, a method of extrapolating laboratory test results for the settlement analysis is needed. The method commonly used, known as the Terzaghi theory of consolidation, is based on several simplifying assumptions. A partial list of these assumptions follows.

1. *Homogeneous Soil*

Homogeneity is assumed in most soil theories, although many soils do not approximate it. Figure I–5 illustrates the nonhomogeneity of a typical sedimentary clay.

2. *Saturated Soil*

Although most naturally deposited clays below the phreatic line approach saturation, earth fills are usually less than 90% saturated.

3. *One-Dimensional Drainage and Compression* [3]

Both drainage and compression are one dimensional in the laboratory test, because the apparatus is constructed to insure it. The extent to which this assumption is not fulfilled in the field can be serious, especially for the compression of clay strata near the point of application of a load which does not extend over a wide area. For example, for a particular airfield fill, the rate of consolidation from both vertical and horizontal drainage was approximately three times what would be expected according to theory based on vertical drainage only. For the soil involved, it was estimated that the horizontal permeability was thirty-four to fifty times the vertical permeability (IX–5).

4. *Constancy of Certain Soil Properties*

The fact that this assumption is seldom complied with is illustrated by Fig. IX–12, which shows that c_v varies widely with pressure for a particular soil. Although so large a variation is not uncommon, it is not as serious as it would at first seem because the pressure increment involved in most practical problems is small.

5. *Straight-Line Plot of Pressure versus Void Ratio*

Not only does the Terzaghi theory assume that a plot of pressure against void ratio at the end of consolidation is a straight line but also that the plot during consolidation is straight. Laboratory tests on many soils have shown that the curve of log pressure against void ratio at the end of the consolidation increment is approximately straight instead of the pressure-void ratio one. For example, see Fig. IX–12. To assume that the plot is straight during the consolidation process means that all compression is due to drainage caused by the hydrostatic excess pressure. This point is discussed further below.

Usually the major source of error in applying the Terzaghi theory is that this assumption is not fulfilled. The error has more effect on predictions of settlement rates than it does on predictions of amounts of settlement.

In Fig. IX–2 is shown a plot of compression versus square root of time.[4] The point d_0 is the start of compression; d_s, the corrected zero point, is found by extending the straight portion of the plot to intersect

---

[2] Basing his report on his studies of buildings founded on Boston blue clay, Fadum (IX–4) reported that the major portion of settlements due to deformation occurred during construction of the buildings. The settlements subsequent to construction were almost entirely due to volume changes within the soil.

[3] A soil specimen can be consolidated under a three-dimensional system of forces in the triaxial compression apparatus. See Chapter XIII.

[4] The square root of time is used rather than time for the abscissa because it is required in one of the fitting methods. See page 85.

the line of zero time; d_{100} is the 100% compression as computed by a fitting method (see page 82); and d_f is the actually measured point of final compression. The compression from d_0 to d_s is called the "initial compression"; that from d_s to d_{100}, "primary compression"; and from d_{100} to d_f, "secondary compression." Although d_0 is usually above d_s, it can also be beneath it, thus indicating a negative initial compression. One contributing factor to a positive initial compression is the compression of entrapped air. Nega-

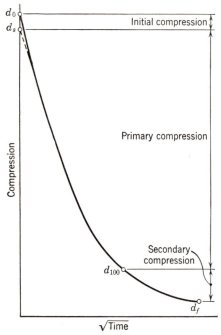

FIGURE IX–2. Compression curve.

tive initial compression is thought to be caused partly by some sort of structural bond (IX–11). Primary compression is due to the drainage of pore water, because of the hydrostatic excess pressure; the secondary compression is probably due to plastic flow or gradual structural adjustment under the imposed load.

The time computations of the Terzaghi theory cover only the primary compression, or consolidation. Therefore, the ratio of primary to total compression or primary compression ratio, r, is an indication of the amount of the total compression covered by the theory. In other words, the larger the value of r, the better the agreement between the computed and the actual rate of settlement in a practical problem.[5]

[5] We should not infer that the primary compression ratio determined from a laboratory test on a small specimen is the ratio that would apply to a soil stratum in nature. The ratio depends on the size of the soil mass undergoing compression.

This short discussion indicates why the Terzaghi theory is not exact; in fact, for some problems it may be only a crude approximation. There are theories available (IX–11) which are more precise, but their complexity restricts their use to a small fraction of the use which the Terzaghi theory receives. In spite of the unfulfilled assumptions in the Terzaghi theory, reasonably good predictions of structure settlements can usually be made from the results of carefully run laboratory tests. Predicted settlements are larger than actual settlements more often than not.[6] Better predictions, naturally, can be made for those cases which have conditions more closely agreeing with the assumptions made in the theory derivation, as, for example, when the soil involved has small secondary effects (inorganic soils appear to have minor secondary effects in comparison with organic soils).

Apparatus and Supplies

Special

1. Consolidation unit
2. Specimen trimmer and accessories
   (a) Miter box
   (b) Wire saw
   (c) Knives
3. Device for placing specimen in container

General

1. Balances (0.01 g sensitivity and 0.1 g sensitivity)
2. Drying oven
3. Desiccator
4. Timer
5. Watch glasses or moisture content cans
6. Scale
7. Evaporating dish

Methods of loading consolidation and shear specimens are discussed in Chapter XI. The two methods most widely used for consolidation tests are the jack with load measurement by platform scales and the wheel or lever system on which weights of known magnitudes are hung. The former method is used in the unit shown in Fig. IX–3, and the latter in the units shown in Fig. IX–4. The apparatus shown in Fig. IX–3 is more portable, and since several people can gather around it simultaneously it has advantages for instructional purposes. That shown in Fig. IX–4 is more compact, and tends to hold the applied loads more nearly constant; it is superior, therefore,

[6] Fadum (IX–4) found that, for the buildings founded on Boston blue clay which he studied, the computed settlements varied from 100% to 200% of the actually observed settlements.

for routine testing, especially where floor space is limited.

FIGURE IX–3. Consolidation unit.

Two types of soil containers, the fixed-ring container and the floating-ring container, are in common use. Diagrammatic sketches of both rings are shown

in Fig. IX–5. In the fixed-ring container, all the specimen movement relative to the container is downward; in the floating-ring container, compression oc-

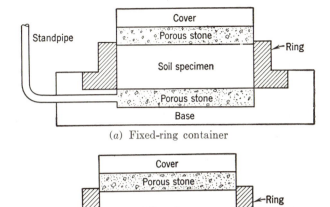

(a) Fixed-ring container

(b) Floating-ring container

FIGURE IX–5. Consolidation specimen containers.

curs toward the middle from both top and bottom. In a test using the floating-ring container, the effect of friction between the container wall and the soil specimen is smaller; on the other hand, only the fixed-ring container can be adapted for permeability tests. The selection of type of loading unit and soil container is

FIGURE IX–4. Frame with consolidation units. (Courtesy of Soil Mechanics Laboratory, Northwestern University, Evanston, Ill.)

based more on personal preference than on any important technical advantages.

There are several types of devices used to trim the test specimen and place it in the container. One consists of a knife edge which fastens to the bottom of the container and cuts the specimen as the container is pushed into the soil. Trimming away the excess soil aids this procedure. Another method is illustrated in Figs. IX–7 and IX–8, and is described in the Recommended Procedure which follows. There is a device which combines the two steps shown in Figs. IX–7 and IX–8; this device lowers the specimen into place as it is trimmed.

Recommended Procedure [7]

Specimen Preparation. Review page 5 before starting the specimen preparation.

1. Measure the height (Z_1 in Fig. IX–6) and diameter of the container; also measure the combined thickness of the upper porous stone and cover, Z_2.

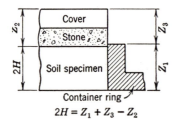

$$2H = Z_1 + Z_3 - Z_2$$

FIGURE IX–6. Dimensions of container.

2. Determine the tare weight, which is the scale beam reading with the empty consolidation apparatus (exclude those parts which lie on top of the specimen) resting on the platform. This tare weight is added to the computed scale loads required to give the desired pressures. If the wheel or lever loading system is used, either check to see that the apparatus is properly counterbalanced or obtain the tare weight.

3. Trim the test specimen, attempting to have the soil strata oriented in the same direction in the consolidation apparatus as they were in nature.[8] First cut a sample of soil approximately 5 in. in diameter and 1¾ in. thick (see page 81 for discussion of specimen size), with one of the faces carefully cut to

[7] This test can be performed better by two or more students. They should be able to run their first test in about 9 hours spread over a period of a week and do the computations in about 6 hours. They will need supervision at the beginning and at the end of the test.

[8] The laboratory test should normally compress the soil in the same direction relative to soil strata as the applied load in the field.

a plane. Place the sample in the trimmer with the plane face down, and trim the corners, rotating the sample a few degrees after each cut until a circular cross section is obtained. See Fig. IX–7.

4. From the soil trimmings obtain at least three representative specimens (of approximately 10 to 15 g weight) for water content determinations.

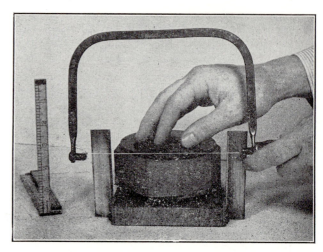

FIGURE IX–7. Trimming a soil specimen.

5. Place the trimmed specimen in the container, using the greatest of care to insure as near a perfect fit as is possible. The specimen can be lowered into the container with the device shown in Fig. IX–8 by rotating the small knob.

FIGURE IX–8. Lowering specimen into container.

6. Trim the top and bottom faces of the specimen flush with the container by means of a wire saw (or knife).

7. Next place the bottom porous stone, which has been soaked in water, on the base of the unit, and then raise the water level above the porous stone.

8. Place the specimen and the container on the porous stone; very carefully place the ring seal, porous stone, and cover on top of the soil specimen. See Fig. IX–9. Next bolt the gutter (at left of Fig. IX–9) to the base.

FIGURE IX–9. Assembling consolidation unit.

Compression Test. In the following procedure, a definite loading program which is satisfactory for most soils is recommended. The reasons back of the choice are discussed later in the chapter.

1. Mount the container with the specimen in the loading unit.

2. Screw the holder with the vertical deflection dial in place and adjust it in such a way that the dial is at the beginning of its release run.

3. Apply the load [9] to give a pressure intensity of $\frac{1}{2}$ kg per cm² [10] on the soil specimen, and start taking time and vertical deflection readings. Compression readings should be taken at total elapsed times [11] of 0, $\frac{1}{4}$, 1, $2\frac{1}{4}$, 4, $6\frac{1}{4}$, 9, $12\frac{1}{4}$, 16, $20\frac{1}{4}$, 25 minutes, etc., until 90% consolidation has been reached. This point can be determined by plotting compression readings versus the square root of elapsed times while the test is in progress, and obtaining 90% consolida-

[9] See the discussion of side friction on pages 80 and 81.

[10] A load of $\frac{1}{4}$ kg per cm² or less may be used for the first increment if it is expected that the early portion of the stress-strain curve will be important. The early portion would be important for a clay which had not been subjected to large loads in its geologic history.

[11] The times suggested for readings give a good spacing of points on the plot used for the square root fitting method (see page 82). Readings taken at other intervals can, of course, also be plotted. More frequent readings, or even a larger test specimen, may be required for soils which compress very rapidly.

tion by the square root fitting method (see page 82). Readings at the predetermined intervals may be stopped at the 90% consolidation point, but occasional observations should be continued until a sufficient number have been taken for the log fitting method (see page 82).

4. At the end of 24 hours, a compression and time reading should be made and then the load increased to 1 kg per cm²; readings should be taken as they were for the 0 to $\frac{1}{2}$ kg per cm² increment.

5. On successive days, loads of 2, 4, and 8 [12] kg per cm² should be applied.

6. After the 8 kg per cm² load has been on for 24 hours, the load is decreased to 2 kg per cm² and then to 0.1 kg per cm². At least 4 hours should be allowed for the 2 and the 0.1 kg per cm² rebound loads; no time readings are normally taken during the rebound.[13]

7. At least two, and preferably more, Z_3 readings should be taken during the test; each Z_3 reading is to be taken at the end of a load increment. See Fig. IX–10. These values are used to compute the thickness of the sample, $2H$, by means of the expression shown in Fig. IX–6.

8. Throughout the test, the container gutter (the gutter can be seen in Fig. IX–10) and the standpipe connection should be kept well filled with water in order to prevent desiccation and to provide water for the rebound expansion.

9. After the final reading has been taken for the 0.1 kg per cm² load, quickly dismantle the apparatus, dry the surface water of the soil specimen, and weigh.

10. Place the weighed specimen in the oven to dry. This enables you to obtain the water content of the whole specimen.

Permeability Determination. The permeability of the consolidation specimen can be computed at each void ratio by means of Eq. IX–6. Also, a direct measurement of the permeability can be made by the constant or variable head permeability test (see Chapter VI). To make a determination by the variable head test, connect a standpipe to the fixed-ring container shown in Fig. IX–5a and follow the procedure given in Chapter VI. The constant head test is run by connecting the water supply (see Fig. VI–3) to the standpipe connection. For constant head tests, an air pressure is often applied to the surface of the water supply

[12] A pressure of 16 kg per cm² or higher may be used if data under larger pressures are desired, and if the apparatus is strong enough to apply such loads.

[13] The unloading procedure suggested in step 6 is arbitrary. If swelling information is required, more complete rebound data are justified.

FIGURE IX–10. Measuring Z_3.

to give a larger head. Both the variable and constant head tests should be run at the end of a load increment when no consolidation is occurring.

Discussion of Procedure

Side Friction. Part of the load applied to a consolidation specimen is transferred to the container wall is $P - 2HF$ and the average load is $P - HF$, where F is the frictional force per height. In the floating-ring, since the soil specimen moves downward at the top and upward at the bottom relative to the ring, the average force is $P - (HF/2)$. The vertical distributions of force are shown for the two rings in Fig. IX–11. We can see, therefore, that the effects of fric-

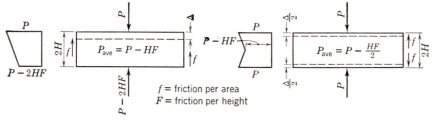

FIGURE IX–11. Side friction.

by friction between the wall and the specimen. Figure IX–11 illustrates the friction effect in both the fixed-ring and the floating-ring containers. In the fixed-ring container, all the soil movement is downward relative to the ring, and thus the entire frictional force is upward on the soil. If a load of P is applied to the top of a specimen $2H$ high, the load at the base tion in reducing the average applied load to the soil are smaller in the floating-ring container than in the fixed-ring one.

To obtain best results from a laboratory test, the magnitude of applied load should be increased in such a way that the average force within the specimen is of the desired amount. However, this is not an easy thing to do because of the difficulty of determining the proper friction value. The friction is a function of the intergranular pressure, and, therefore, varies during the consolidation process; it also depends on the size of the soil container. More important, the frictional characteristics are likely to be different for different soils.

Taylor's work (IX–11) indicates the magnitude of the frictional force on Boston blue clay. He reports that the total frictional force ($2HF$) ran from 12% to 22% of the applied load, P, for the remolded clay and from 10% to 15% of P for the undisturbed clay.

Since the friction makes the effective loads smaller than the applied loads, to neglect it makes the pressure–void ratio curve fall off the true one in the direction of larger pressures. Taylor (IX–11), however, reports that for Boston blue clay side friction did not appreciably affect either the coefficient of consolidation or the coefficient of compressibility. Side friction is normally neglected in routine consolidation testing because of its minor effects and because of the difficulty of determining its magnitude. The suggestion has been made (IX–1), however, that to increase the applied load by 10% would probably compensate for friction in many soils.

Specimen Size. The size of a consolidation test specimen is important for several reasons. Those which point to smaller specimens are:

1. *Economics*

Generally, the cost of sampling operations increases rapidly as the diameter of the soil sample recovered is increased. A 4½ or 4¾ in. sample is necessary to obtain the 4¼ in. diameter specimen often used in consolidation tests. The cost of taking such large samples is usually prohibitive.

2. *Consolidation time*

The thinner the specimen, the smaller the distance the escaping pore water must flow and, therefore, the shorter the time required for consolidation. Equation IX–2 indicates that the time for consolidation varies as the square of the specimen thickness.

3. *Side friction*

The thinner the specimen for a given diameter, the less the side friction; this is true because of the smaller lateral surface area in contact with the wall of the container.

On the other hand, a consideration of major importance which points to the use of larger specimens is the disturbance to the soil structure which occurs during specimen preparation. Van Zelst (IX–12) has presented data which suggest that the thickness of the disturbance zone caused by the trimming is essentially independent of the specimen thickness. He found that, for one particular clay, a zone of 0.1 in. at each of the faces of the specimen was remolded by trimming. This disturbance tended to lower the resulting pressure–void ratio curve (see Fig. IX–12). Since the zone of disturbance appears to be independent of specimen thickness, the thinner the specimen the more important the effect of disturbance.

As would be expected, the results of laboratory tests depend on the size of the specimen employed. A series of tests (IX–9) on five widely different clays in which both 4¼ in. diameter by 1¼ in. thickness and 2¾ in. diameter by 0.85 in. specimens were used indicated that the pressure–void ratio curve was essentially independent of size. Unfortunately, the rate of compression was greatly dependent on size; higher coefficients of consolidation were obtained on the larger specimens.

In view of the foregoing considerations, a ratio of specimen diameter to thickness of about three to four is recommended. Diameters greater than 2½ to 2¾ in. are desirable.

Loading Procedure. The procedure recommended in the preceding pages of doubling the applied load every 24 hours is generally accepted. The dependence of the coefficient of consolidation on $(P_2 - P_1)/P_1$, sometimes called "load-increment ratio," has been shown (IX–11) to be of much consequence for remolded Boston blue clay. In this soil, increasing the load-increment ratio from one, which corresponds to doubling the applied load, to two caused an increase in the coefficient of consolidation of approximately 30%. Ratios less than one may cause the plastic properties of the soil to have an appreciable effect on the test (IX–7).

For some clays a new load can be applied every 12 hours without greatly affecting the results. Although this procedure cuts the testing time almost in half, it is not usually convenient because of the night readings required. To load a specimen more rapidly appears to have a pronounced effect on the results because of the long time required for the plastic flow in secondary compression. Keeping the elapsed times between load applications constant for the duration of the entire test is necessary for consistent results.

In a program involving many consolidation tests on the same type of clay, comparative tests can be used to determine whether or not loads can be of short duration. Often it may be found satisfactory to load a soil every 3 or 4 hours for the small loads, and then every 24 hours for the stages of the test for which good results are necessary.

Calculations

Probably the easiest way to present the calculations for the consolidation test is to explain how each column of the data sheet for the Numerical Example (page 86) was obtained. The columns are lettered and will be referred to by letter.

Column a. In this column are listed the pressures to which the specimen is to be subjected.

Column b. The figures in this column were obtained by adding the tare weight to the product of the specimen area and the desired pressure intensity. Any increment to be added for friction should be included here.

Column c. Here are recorded the final vertical dial readings for each load.

Column d. The change in final dial readings is the compression for a load increment, and is found by subtracting successive numbers in column c.

Column e. A list of the measured Z_3 readings.

Column f. The sample thickness, $2H$, is computed for each Z_3 reading from the expression

$$2H = Z_1 - Z_2 + Z_3 \quad \text{(see Fig. IX–6)}$$

Column g. All the calculated $2H$ values in column f are reduced to a common load by means of the Δ dial values from column d. The average of the $2H$ values is written in column g and underlined. The other values of $2H$ are computed from this average, using the Δ dial values in column d.

Column h. The void height is found by subtracting the height of solids, $2H_0$, from the specimen height, $2H$, read from column g. The value of $2H_0$ is found from

$$2H_0 = \frac{W_s}{G\gamma_w A} \quad \text{(IX–1)}$$

Column i. The void ratio, e, is equal to the void height divided by the height of solids. This is true since the area A can be canceled from the following expression:

$$e = \frac{V_v}{V_s} = \frac{A(2H - 2H_0)}{A 2H_0}$$

Column j. In this column are listed the times for 90% of primary compression if the square root fitting method is used, or the times for 50% of primary compression if the log fitting method is used. Both the fitting methods are explained below.

Column k. The coefficient of consolidation, c_v, is computed from the following equations: [14]

1. Square Root Fitting Method:

$$c_v = \frac{0.848H^2}{t_{90}} \quad \text{(IX–2a)}$$

2. Log Fitting Method:

$$c_v = \frac{0.197H^2}{t_{50}} \quad \text{(IX–2b)}$$

In Eqs. IX–2a and IX–2b, H is the average thickness per drainage surface for the load increment. H for any increment is found by dividing the sum of the $2H$ values (column g) for the boundary loads of the increment by four. Values t_{90} and t_{50} come from column j.

Fitting Methods

Square Root Method (see left plot on page 85). Make a plot of compression dial reading against the square root of elapsed time. Extend the straight-line portion of the curve back to intersect zero time and obtain the corrected zero point, d_s. Through d_s draw a straight line having an inverse slope 1.15 times the tangent. This straight line cuts the compression-time curve at 90% compression.

Log Method (see right plot on page 85). Plot compression dial readings against the log of time. The two straight portions of the curve are extended to intersect at 100% primary compression. The corrected zero point, d_s, is located [15] by laying off above a point in the neighborhood of 0.1 minute a distance equal to the vertical distance between this point and one at a time which is four times greater. The 50% compression point is halfway between the corrected zero point and the 100% compression point.

The primary compression ratio, r, can be found from the following equations:

1. Square Root Fitting Method:

$$r = \frac{\frac{10}{9}(d_s - d_{90})}{d_0 - d_f} \quad \text{(IX–3a)}$$

2. Log Fitting Method:

$$r = \frac{d_s - d_{100}}{d_0 - d_f} \quad \text{(IX–3b)}$$

[14] See reference IX–10 for derivation of Eqs. IX–2a and IX–2b.

[15] Since a log scale is used on the abscissa, zero time cannot be plotted. The method of locating it assumes that the early portion of the curve is a parabola.

in which d_s = corrected zero point,

d_{90} = compression dial reading at 90% primary compression by square root fitting method,

d_{100} = compression dial reading at 100% primary compression by log fitting method,

d_0 = compression dial reading at zero time,

d_f = final dial reading.

From the calculated void ratios, a plot of void ratio, e, vs. log of pressure, p, can be plotted. The slope of this curve is called the compression index, C_c, or

$$C_c = -\frac{de}{d(\log_{10} p)} \qquad \text{(IX–4)}$$

The slope of the pressure–void ratio plot is called the coefficient of compressibility, a_v. Since the log p vs. e curve is usually plotted rather than the p vs. e curve,[16] a_v can be found from C_c by

$$a_v = \frac{0.435 C_c}{p} \qquad \text{(IX–5)}$$

in which p is the average pressure for the increment, i.e., $(p_1 + p_2)/2$ is used for p.

Having the values of a_v (Eq. IX–5) and c_v (Eq. IX–2), we can compute the permeability from

$$k = \frac{c_v a_v \gamma_w}{1 + e} \qquad \text{(IX–6)}$$

in which γ_w = the unit weight of water.

The maximum intergranular pressure to which the clay has been previously consolidated, or $\bar{\sigma}_c$, can be approximated (IX–2) from the void ratio–log of pressure plot by the following procedure (see Fig. IX–12):

1. Locate on the plot the point of maximum curvature.
2. At this point, draw a horizontal line and a tangent to the curve.
3. Bisect the angle formed by the horizontal and tangent lines.
4. Intersect the bisector by extending back the straight portion of the log p vs. e curve which exists at the higher pressures.
5. The pressure at the intersection of the bisector and the tangent produced backwards is the approximation of the maximum precompression pressure.

[16] The log p vs. e curve is more nearly straight and its slope, therefore, is easier to determine.

Results

Method of Presentation. The results of a consolidation test are often presented by plotting [17] on the same sheet the log of pressure against the following: (1) void ratio, e; (2) coefficient of consolidation, c_v; and (3) primary compression ratio, r. (See Fig. IX–12.)

Typical Values. The fitting method which is better for a given clay must be found by trial. On some clays one will work better than the other; on many clays both methods work. The square root method is more convenient because it permits the determination of primary compression as it is completed, through the plotting of observations as they are made. Thus it can be determined for how long a time frequent readings must be taken.

The slope of the void ratio–log of pressure curve, or compression index, C_c, is a measure of the compressibility of the clay. That portion of the curve prior to the maximum past consolidation pressure, or the recompression portion, usually has a much flatter slope than the portion for pressures greater than the maximum past consolidation pressure, or "virgin" portion of the curve. Thus a clay [18] may be classified as "stiff" for applied loads on the recompression curve and as "compressible" for applied loads on the virgin curve. Many consolidation tests on many different clays have shown that the virgin curve is nearly always close to a straight line.[19]

The rate of consolidation under a load increment is represented by the coefficient of consolidation, c_v. Since the coefficient of consolidation is a function of permeability, void ratio, and compression index, it is a function of applied pressure. The structure of the clay influences the value of c_v; e.g., remolded Boston blue clay has a value of c_v which is only a third of the undisturbed one.

[17] Although a definite void ratio exists for a pressure, the coefficient of consolidation and primary compression ratio exist for an increment of pressure. Void ratio, therefore, is plotted against the pressure at which it occurs, and the coefficient of consolidation and primary compression ratio are each plotted against the average pressure of the corresponding loading increment. (See Fig. IX–12.) Read the section entitled Typical Values for information concerning which fitting method should be used in computing the coefficient of consolidation and the primary compression ratio.

[18] A soil existing at the largest intergranular pressure to which it has ever been consolidated is called a normally consolidated soil; on the other hand, a soil existing at an intergranular pressure smaller than the maximum to which it has been consolidated is a precompressed or preconsolidated soil.

[19] The plot of log e vs. log p may be nearer a straight line for very compressible soils, especially organic soils such as peat.

An indication of the degree to which consolidation characteristics can vary among different soils can be obtained from the list of characteristics [20] for several widely different soils given in Table IX–1.

TABLE IX–1

	Virgin Compression Index *	Coefficient of Consolidation † in 10^{-4} cm²/second
Mexico City clay (volcanic origin; composed mostly of montmorillonite)	4.5	0.2 to 2.5
Boston blue clay (marine deposit of glacial clay, composed partially of illite)		
Undisturbed	0.22	10 to 20
Remolded	0.26	1 to 6
Morganza Louisiana clay (fluvial deposit; composed mostly of illite)	0.44	0.5 to 1.0
Newfoundland peat	8.5	0.2 to 3
Maine clay (silty, glacial clay composed partially of illite)	0.5	20 to 40

* Determined from plot of void ratio vs. log pressure where the pressure was in units of kg/cm².

† For virgin portion of compression curve.

Discussion. In the preceding pages several of the variables which influence the consolidation test have been discussed. There is another, however, which may be important but is seldom considered. This is temperature. Since temperature affects permeability (see Eq. VI–3) and permeability affects the coefficient of consolidation (see Eq. IX–6), temperature must affect the coefficient of consolidation. For example, Eqs. VI–3 and IX–6 can be used to show that the coefficient of consolidation at 95° F is 35% greater than that at 70° F; both 95° F and 70° F are reasonable for a laboratory.

Gray (IX–6) reported that an increase in temperature appeared to have no influence on the amount of primary compression but caused a measurable increase in secondary compression. Since the secondary compression is considerably more important in organic soils, temperature effects are probably more serious on the compression curves of organic soils. The explanation of the influence of temperature on the plastic properties of clays is not known.

The effect of temperature is difficult to account for because of the incomplete understanding of its influence coupled with the fact that it may vary during a

[20] These values are based on tests on three to five specimens of each different clay.

test. Also data on the temperature of the soil as it exists in the ground would be necessary in applying the test results directly. Test temperatures should be recorded, however, to help explain any irregularities in test results and to aid at least a semi-quantitative consideration of temperature in the in situ soil.

Numerical Example

The data on pages 85–87 are from a consolidation test on a glacial silty clay from Maine. Figure IX–12 indicates a maximum precompression of 2.4 kg per cm², which is in approximate agreement with what would be expected from the estimated weight of the glacier that once rested on this soil.

REFERENCES

1. Burrows, R. E., "An Experimental Study of Side Friction in the Consolidation Test," Master of Science Thesis, Department of Civil and Sanitary Engineering, Massachusetts Institute of Technology, 1948.
2. Casagrande, A., "The Determination of the Pre-Consolidation Load and Its Practical Significance," *Proceedings International Conference on Soil Mechanics and Foundation Engineering*, Vol. III, pp. 60–64, 1936.
3. Enkeboll, William, "Investigation of Consolidation and Structural Plasticity of Clay," Doctor of Science Thesis, Department of Civil and Sanitary Engineering, Massachusetts Institute of Technology, 1946.
4. Fadum, R. E., "Observations and Analysis of Building Settlements in Boston," Doctor of Science Thesis, Graduate School of Engineering, Harvard University, 1941.
5. Gould, J. P., "Analysis of Pore Pressure and Settlement Observations at Logan International Airport," Doctor of Science Thesis, Graduate School of Engineering, Harvard University, December, 1949.
6. Gray, Hamilton, "Research on the Consolidation of Fine-Grained Soils," Doctor of Science Thesis, Graduate School of Engineering, Harvard University, April, 1938.
7. Marsal, Raul J., "Investigation of Consolidation and Plastic Resistance on Clays," Master of Science Thesis, Massachusetts Institute of Technology, Department of Civil and Sanitary Engineering, 1944.
8. Rutledge, P. C., "Relation of Undisturbed Sampling to Laboratory Testing," *Transactions American Society of Civil Engineering*, Vol. 71, p. 1211, 1944.
9. Starnes, W. L., and E. E. Bennett, "Sample-Size Requirements for Consolidation Tests," Master of Science Thesis, Department of Civil and Sanitary Engineering, Massachusetts Institute of Technology, 1947.
10. Taylor, D. W., *Fundamentals of Soil Mechanics*, John Wiley and Sons, New York, 1948.
11. Taylor, D. W., "Research on the Consolidation of Clays," Department of Civil and Sanitary Engineering, Massachusetts Institute of Technology, Serial No. 82, 1942.
12. Van Zelst, T. W., "An Investigation of the Factors Affecting Laboratory Consolidation of Clay," *Proceedings Second International Conference on Soil Mechanics*, Vol. VII, Paper No. II c 4, p. 52, 1948.

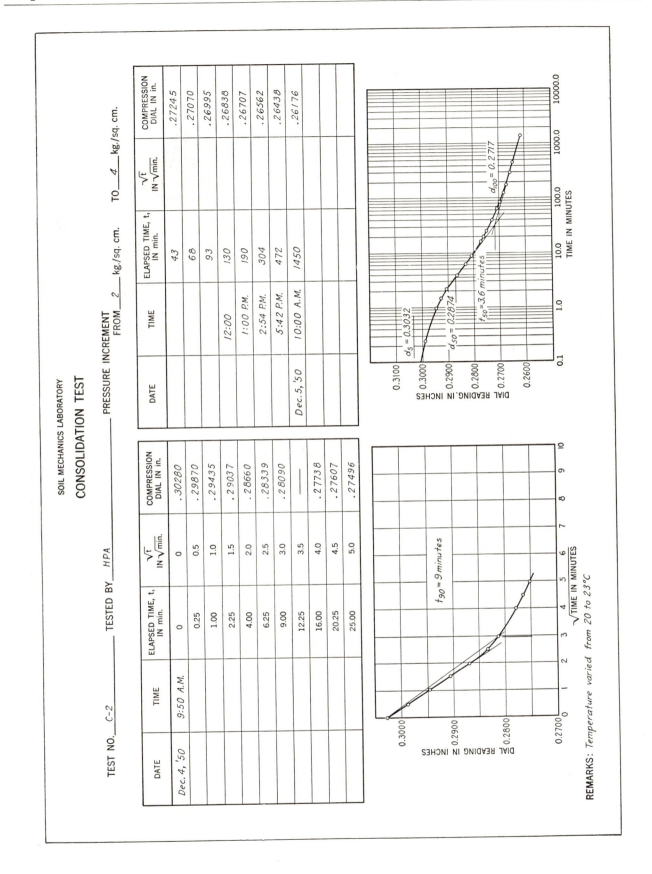

SOIL MECHANICS LABORATORY

CONSOLIDATION TEST

SOIL SAMPLE _Sandy silty clay; gray; inorganic; glacial origin; sedimentary deposit; extremely sensitive; medium plastic; soft when remolded._

LOCATION _Union Falls, Maine_
BORING NO. _Pit-A-3_   SAMPLE DEPTH _3_
SAMPLE NO. _Pit-A-3_
SPECIFIC GRAVITY, G_s, _2.83_

APPARATUS MEASUREMENTS
CONTAINER HEIGHT, Z_1, _3.321_ cm. _1.263_ in.
CONTAINER DIAMETER _10.8_ cm. ___ in.
CONTAINER AREA, A, IN sq. cm. _91.8_
STONE + COVER
THICKNESS, Z_2, _2.15_ cm. _0.847_ in.

SOLIDS HEIGHT, $2H_0 = \dfrac{W_s}{G_s \gamma_w A}$ = _1.71_ cm. _0.675_ in.

APPLIED LOADS
SIDE FRICTION ALLOWANCE IN % _0_
1 kg./sq. cm. = _202_ lbs. SCALE LOAD
TARE IN lbs. _4.4_

TEST NO. _C-2_
DATE _Nov.30,1950_
TESTED BY _HPA_

DEGREE OF SATURATION IN %
TEST START _85.2_
TEST END _100.0_

REMARKS:
Sample had fine silt lenses running in horizontal direction.

WATER CONTENT

SPECIMEN LOCATION	BEGINNING OF TEST			END OF TEST
	TOP	BOTTOM	SIDE	ENTIRE
CONTAINER NO.	F4	F29	F9	B11
WT. CONTAINER + WET SOIL IN g	23.900	28.168	24.421	704.7
WT. CONTAINER + DRY SOIL IN g	21.740	24.800	22.230	592.5
WT. WATER, W_w, IN g	2.160	3.368	2.191	112.2
WT. CONTAINER IN g	13.184	13.294	13.146	147.9
WT. DRY SOIL, W_s, IN g	8.556	11.506	9.084	444.6
WATER CONTENT, w, IN %	25.2	29.3	24.1	25.2

a	b	c	d	e	f	g	h	i	j		k	
APPLIED PRESSURE IN kg./sq. cm.	SCALE LOAD IN lbs.	FINAL DIAL IN in.	DIAL CHANGE IN in.	Z_3 IN in.	SPECIMEN HEIGHT $2H = Z_1 - Z_2 + Z_3$ IN in.	2H FROM DIAL CHANGE IN in.	VOID HEIGHT, $2H - 2H_0$ IN in.	VOID RATIO, $e = \dfrac{2H - 2H_0}{2H_0}$	FITTING TIME IN sec.		COEF. OF CONSOL., c_v, IN sq. cm./sec.	
									t_{90}	t_{50}	$.848H^2/t_{90}$	$.197H^2/t_{50}$
0	44.0	.34000				1.262	0.587	0.870				
¼	94.5	.33287	.00713			1.255	0.580	0.860	318	59	.00683	.00879
½	145.0	.32860	.00427			1.250	0.575	0.852	276	66	.00777	.00752
1	246.0	.32024	.00836	0.825	1.241	1.242	0.567	0.840	284	58	.00725	.00860
2	448.0	.30280	.01744			1.225	0.550	0.815	264	95	.00790	.00510
4	852.0	.26176	.04104	0.768	1.184	1.184	0.509	0.755	540	216	.00368	.00214
8	1660.0	.20762	.05414			1.129	0.454	0.673	516	201	.00354	.00212
2	448.0	.21179	.00417			1.133	0.458	0.679				
0.1	64.2	.23395	.02216			1.155	0.480	0.711				

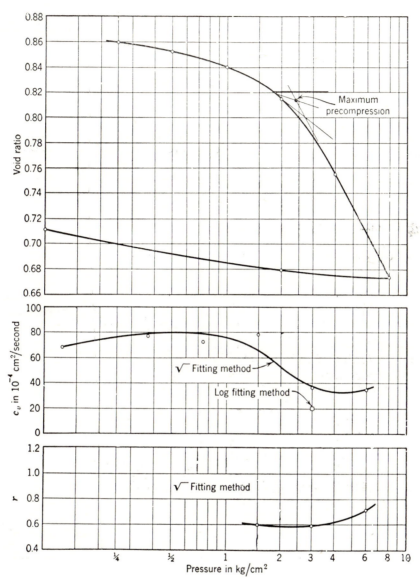

FIGURE IX–12. Consolidation test.

CHAPTER
X

Direct Shear Test on Cohesionless Soil

Introduction

In all soil stability problems, such as the design of foundations, retaining walls, and embankments, knowledge of the strength of the soil involved is required. The determination of the proper strength to use in a stability problem, especially one involving cohesive soil, can be the most difficult type of question which arises in soil engineering. This chapter and the following four are devoted to the perplexing subject of laboratory strength testing of soils. A brief discussion of strength theory is given for cohesionless soil in this chapter and for cohesive soil in Chapter XII. The reader is referred to reference X–6 for a more complete presentation of this theory.

Strength Theory. Depending on the source of its strength, a soil can be placed in one of two groups, namely, cohesionless and cohesive. As their names imply, cohesionless soils are soils which have no cohesion, or attraction, between individual particles and cohesive soils are soils the individual particles of which exhibit interattraction.

The resistance to shear of a cohesionless soil is derived from friction between grains and the interlocking of grains. Friction between soil grains is similar to friction between any surfaces, as, for example, between the two blocks shown in Fig. X–1. When the top block is slid along the bottom block, a shear force is applied to the surface of the bottom block which is equal to the normal force acting between the blocks multiplied by a coefficient called the "coefficient of friction." In soils, the friction may be either sliding friction, as between the two blocks in Fig. X–1, or rolling friction. For example, if a large enough shear force were applied to soil grain A (Fig. X–2), it could be moved to position B by either sliding or rolling, or

a combination of the two. Normally, no attempt is made to distinguish between sliding friction and rolling friction.

To move the soil grain from position B to position C (Fig. X–2) against the applied normal force requires

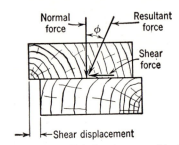

FIGURE X–1. Friction between blocks.

work equal to the distance d times the normal force. This work is the quantitative measure of the phenomenon termed interlocking. Because interlocking

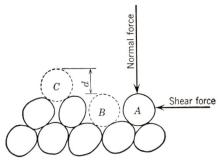

FIGURE X–2. Friction in soil.

occurs to a greater extent when soil grains are closer together, dense soils show a higher shear strength at small shear displacements than loose soils. During the initial part of shear on a dense sand the volume of the soil increases, with the result that the effects of interlocking are reduced.

Figure X–3 shows an element of soil being subjected to a shear force, which is equal to the normal force times tan α. If the area of the potential shear surface is A, the shear stress, τ, is equal to the shear force divided by A; and the normal stress, σ, is equal to the normal force divided by A. It follows that

$$\tau = \sigma \tan \alpha$$

The shear strength, s, is the shear stress which is necessary to cause slippage on a surface through the soil; or

$$s = \sigma \tan \phi$$

where ϕ is the angle α at slippage. The shear strength of the soil can be expressed as s ("shear strength") or

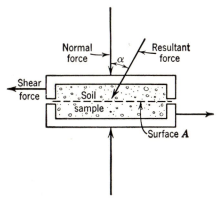

FIGURE X–3. Shear of soil in a box.

tan ϕ ("coefficient of friction") or ϕ ("angle of internal friction" or "friction angle").

Because of the phenomenon of interlocking described above, the strength of a dense cohesionless soil tends to be greater at small displacements than at large displacements, where the effects of interlocking have been overcome. The higher strength is called "maximum" or "peak" strength; the lower strength is "ultimate" strength.[1] The relative magnitudes of these two strengths are discussed later.

In cohesionless soil the volume changes which occur during shear can be more important than the actual value of the angle of internal friction. If a saturated cohesionless soil shears more rapidly than its pore water can flow in or out, normal stresses are thrown into the water, which has no shear resistance. Thus

a loose saturated cohesionless soil which contracts during shear may lose its strength, or liquefy, as the stress acting normal to the shear plane is carried by water and does not generate friction.

This importance of volume changes led to the development of the concept of "critical void ratio" (X–2). The critical void ratio, e_c, of a soil is the void ratio that exists prior to a shearing process in which the net volume change at failure is zero. In other words, a soil at its critical void ratio will have the same void ratio at failure that it had prior to shear. If a soil denser than its critical void ratio were sheared, it would exhibit a net expansion at failure; therefore, no normal compressive stresses are transferred to the pore water, and there will not be any loss of strength[2] or liquefaction.

Unfortunately, critical void ratio is not a constant property of the soil, but depends on the details of the test procedure from which it is obtained. Also, knowledge of critical void ratios of in situ soils is limited.[3] A particular critical void ratio, however, as defined on page 105, is a valuable guide for a comparative study of several soils. There is evidence that tests on small specimens of a soil indicate an e_c which is larger than that of the in situ soil. A. Casagrande (X–9) suggests that tests on large specimens may give an e_c closer to the one which the soil has in nature.

Methods of Shear. The three common methods of shear testing are direct shear, cylindrical, or triaxial, compression, and torsional shear. In a direct shear test the soil is stressed to failure by moving one part of the soil container relative to another. This type of shear is illustrated in Fig. X–3. When a shear force of sufficient magnitude is applied, the top of the box moves relative to the bottom, causing the soil to shear along surface A. Different types of containers and other details of direct shear testing are presented in the following pages of this chapter.

In torsional shear a circular column of soil is subjected to a twisting moment as shown in Fig. X–4. The moment is normally applied through a disk at the top or bottom; the disk usually has ribs to help prevent slippage between it and the soil. If desired,

[1] As to whether the peak or ultimate strength should be used in an actual problem depends on the particular problem. The peak strength is used more frequently than the ultimate. When the term "strength" is used in this book, the peak strength is meant unless otherwise stated.

[2] In fact, if the rate of shear is too rapid to permit the free inflow of water when the specimen tends to expand, the soil has a greater strength, because the tension thrown into the water causes an equal increase in intergranular compression.

[3] See page 417 and following in "Notes on the Design of Earth Dams," by A. Casagrande, in Vol. 37, No. 4, October, 1950, issue of the *Journal of the Boston Society of Civil Engineers* for a discussion of critical void ratio in situ soils. Casagrande points out that susceptibility of a soil in situ to liquefaction may depend on the amount and the rate of applied shear as well as the void ratio.

a lateral pressure can be applied to the specimen being tested.

The cylindrical compression test, also called the triaxial test, consists of axially loading a cylinder of soil to induce failure. Figure X–5 is a sketch of a soil specimen being tested in cylindrical compression. Normally the soil specimen is encased in a rubber membrane, and a uniform pressure is applied around it by a fluid.

The main advantage of the torsional shear test over the other two types is that the cross section of

FIGURE X–4. Torsional shear.

the soil remains more nearly constant during shear. In both the other tests, the sample is often badly distorted at ultimate failure. This distortion causes nonuniform stresses and strains within the soil, and often makes it difficult to measure accurately the effective area of the failure surface. The most dependable measure of the ultimate shear strength of a soil, therefore, can probably be obtained from torsional shear tests. This advantage, however, is more than outweighed by the fact that the shear displace-

FIGURE X–5. Cylindrical compression.

ments vary as the specimen radius, thus exaggerating progressive failure.[4] This effect is reduced somewhat if an annular-shaped soil specimen is used rather than a solid cylinder.

Nevertheless, the direct shear and certainly the triaxial test are superior to the torsional test for normal laboratory testing. Torsional shear is more common in Europe than in the United States. In fact, at the present time, almost no use is made in the United

[4] Progressive failure is discussed on pages 98 and 99.

States of torsional shear testing in soils.[5] Although the direct shear test is widely used in this country, it is not gaining in popularity as is the triaxial test. The reasons for this trend are discussed in Chapter XI.

Apparatus and Supplies

Special

  1. Direct shear machine

General

  1. Tamper for compacting soil
  2. Balance (0.1 g sensitivity)
  3. Drying oven
  4. Calipers
  5. Straight edge
  6. Large evaporating dish
  7. Spoons, etc.
  8. Timer

The direct shear machine consists of two main parts, the soil container and the loading unit. There

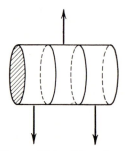

FIGURE X–6. Ring shear.

are several types of shear machines in common use; their names are derived from the type of soil container and loading unit used. A two-piece square box (see Fig. X–3) is one type of soil container. Another type consists of rings which can be subjected to either single or double shear. (See Fig. X–6.) A normal load can be applied to the ends of the specimen soil.[6] A third is illustrated by the specimen in Fig. X–7, which is failed by applying the shear forces as shown. Lateral pressures also can be applied to this specimen.

A soil specimen in the third container (Fig. X–7) undergoes more uniform strains and maintains a more

[5] Hvorslev has constructed a torsional shear device at the Waterways Experiment Station, Vicksburg. This device tests an annular specimen of 2.5 in. inside diameter, 4.5 in. outside diameter, and approximately 0.75 in. high. This apparatus may prove valuable for slow tests (see page 111) on clay because time required for drainage would be less than in the triaxial test.

[6] For more information on the shear of soil contained in rings see references X–1, X–3, and X–4. Reference X–3 also describes an automatic recording device which can be employed for either stress- or strain-controlled tests.

nearly constant area during shear than in the others. However, its complexity and the difficulty of specimen preparation offset this advantage. Specimen prepara-

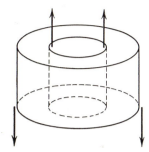

FIGURE X-7. Direct shear of annular specimen.

tion can be very easy for the ring container (Fig. X-6). If the rings are placed within the sampler (see page 5), the sampling operation leaves the soil in the rings ready for testing. This feature is probably more than offset by the better control that can be ob-

apart at a given rate and measuring the resulting force. Units which follow the first procedure are called stress-controlled loading units and those that follow the second are called strain-controlled loading units. Results of tests using the two types of load application have agreed well (X–7). Both types of units are widely used.

Stress-controlled units are preferable for running shear tests at a very low rate, because, with this type of unit, an applied load can more easily be kept constant for any given period of time. In addition, loads can be more conveniently applied and removed rapidly. The ultimate shear resistance cannot be obtained readily [7] in a stress-controlled unit, however, because of the excessive displacement imposed after the maximum resistance has been exceeded. The strain-controlled units, therefore, have an advantage over the stress-controlled ones since the ultimate resistance and a better measure of the peak resistance

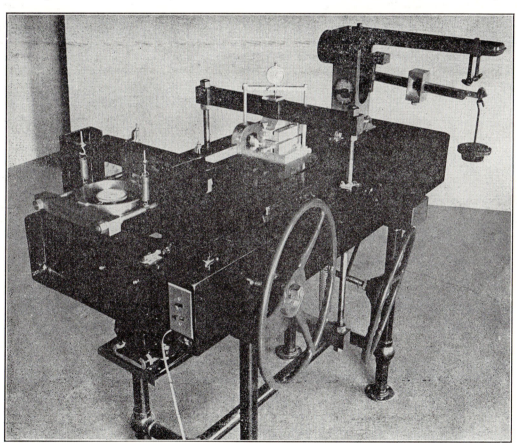

FIGURE X-8. Strain-controlled direct shear machine.

tained with the box. The box enjoys the widest acceptance in the United States.

The shear force can be applied either by increasing the force at a given rate and measuring the resulting displacements, or by moving the parts of the container

can be obtained. A strain-controlled test is a little easier to perform, since the rate of loading in stress control must usually be manually regulated.

[7] A skilled technician can obtain a measure of the ultimate resistance of the soil by decreasing the applied load.

In Fig. X–8 is shown a strain-controlled box-shear machine. The motor moves the lower part of the shear box, which is fastened to the platform; the upper half of the box is held by the horizontal yoke. The shear force is measured by the proving ring [8] attached to the yoke, and the normal load is measured by the platform scales. The shear box of this machine

3. Put the parts of the soil container together and attach it to the shear machine.

4. Weigh a dish of the dry [10] cohesionless soil which is to be tested.

5. Place the soil in a smooth layer approximately $\frac{1}{2}$ in. thick.[11] If a dense sample is desired, tamp the soil.

(a)

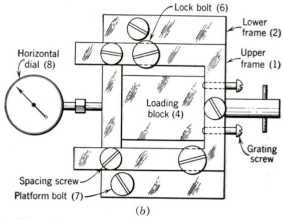
(b)

FIGURE X–9. Shear box.

can be equipped with either toothed gratings or porous stones. The use of gratings gives little greater peak strength [probably around 1% (X–6)] because they transmit the load to the soil better.

Recommended Procedure [9]

The detailed procedure for a direct shear test depends, of course, on the type of apparatus used. The following procedure is presented in such a way that the first part of each step is general and the last part detailed. The general part will serve as a guide for tests on most shear apparatus; the detailed portion, which is in parentheses, applies to the strain-controlled machine shown in Fig. X–8. Figures X–9 and X–10 will be referred to in the detailed procedure.

1. Measure the soil container (L, b, and x in Fig. X–10).

2. Either counterbalance the device used to apply the normal load or obtain the tare weight. The tare weight is the scale reading when the normal load is zero.

6. Reweigh the soil and dish. The difference between this weight and the previously determined weight is the amount of soil used.

7. Make the surface of the soil level. (Use a leveler, like the one shown in Fig. X–10.)

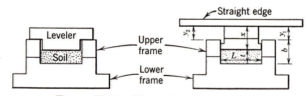

FIGURE X–10. Dimensions of shear box.

8. Put the upper grating, or stone, and loading block on top of the soil.

9. Measure [12] the thickness of the soil specimen. In Fig. X–10,

[10] The soil should be air-dried and not oven-dried; or, if it has been dried in the oven, it should be allowed to come into moisture equilibrium with the atmosphere. There is some indication (X–8) that oven drying may alter the angle of internal friction. This point is of little importance for soils in nature because they would hardly exist in an oven-dry state.

[11] See the discussion of specimen thickness on page 93.

[12] A more accurate measurement of soil volume can be obtained if this measurement is made after the normal load has been applied, but this procedure is more difficult and only a small difference is involved.

[8] In Appendix A, pages 148–151, the design, use, and calibration of proving rings are discussed.

[9] This test can be run better by two or more students. Performing their first test, they should be able to make three runs in about 2 hours and do the computations in about 2 hours. They will need supervision for their first run.

$$t = b + y_a - x$$

in which y_a = the average of y readings on all four sides,

x = the combined thickness of the loading block and upper grating,

b = inside depth of the shear box.

10. Apply the desired normal load.

11. Separate [13] the two parts of the soil container. (Remove the vertical lock screws and raise the upper frame by turning the spacing screws. Tighten the two horizontal grating screws on the upper frame and then turn the spacing screws clear of the lower frame.)

12. Attach the extensometers which measure shear and normal displacements. (Adjust the dial which measures normal displacement to read either expansions or contractions.) Record the initial readings on all dials.

13. Before proceeding with the test, carefully check to see that there is no connection between the two parts of the soil container except the soil.

14. Start the loading. Take readings of shear force, time, and shear and normal displacements.

15. For a strain-controlled test, take a set [14] of readings every 15 seconds for the first 2 minutes, and then a set every 0.03 in. of horizontal displacement. Continue the test to a horizontal displacement of approximately 15% of the length of the specimen unless a constant shear force is obtained first.

15a. For a stress-controlled test, take a set of readings before each new load is added. Continue the test to failure.

Discussion of Procedure

Spacing of Soil Container Parts. For most soils, a spacing of approximately 0.04 in. between the two halves of the soil container is satisfactory. Actually the choice of spacing is dependent on the size of the largest soil grain and the denseness of the soil which is being tested. The parts of the box should be farther apart than the diameter of the largest particle to prevent the top half from riding up on a grain which gets between the edges. Since a loose soil may compress enough to cause the halves of the container to scrape, a larger spacing is required for loose soils. Too close a spacing is indicated by a snapping and crushing of grains, accompanied by jerky readings.

[13] See discussion of spacing in a following paragraph.

[14] The frequency of readings depends on the particular test at hand. In order to aid the student, a frequency of readings is suggested for all the shear tests. These suggested frequencies will usually give more data than are needed to analyze the test properly. The student should compute and plot only the data he finds he needs.

Rate of Strain. A rate of shearing displacement of approximately 0.05 in. per minute is often used. Tests (X–7) on Ottawa sand have shown that within the range investigated, which was 0.1 to 0.006 in. per minute, the effect of rate of strain on the friction angle was always less than 2%. Within reasonable limits, then, the rate at which a cohesionless soil is sheared is not important. However, as pointed out on page 89, rapid shear of a saturated soil may throw stresses into the pore water, thereby causing a decrease in the strength of a loose soil or an increase in the strength of a dense soil. On page 104, the effect of very rapid shear on dry sand is discussed.

In the stress-controlled test, increments of shear force can be added either at regular intervals of time or after displacement has ceased under the existing force. A load increment equal to some percentage— e.g., 10%—of the estimated shear strength can be used.

Size of Shear Box. The larger the maximum size particle which a soil contains, the larger the shear box needed. The box of the machine shown in Fig. X–8 is 3 in. by 3 in. This size, and the size 4 in. by 4 in. are both common. Because the use of larger boxes requires larger loads to obtain the same pressures, very large boxes are not usually practical. Tests (X–7) on Ottawa sand in which both a 3 in. by 3 in. box and a 12 in. by 12 in. box were used indicated that the size of the box had a negligible effect on the results.

Thickness of Soil Specimen. The procedure above calls for a specimen thickness close to 0.5 in. Within limits, the thicker the specimen used, the more the progressive action, and, therefore, the lower the maximum shear strength. Tests (X–7) on Ottawa sand showed that a peak strength approximately 3% lower resulted from using a sample thickness of 0.75 in.

Moisture Content of Soil. As noted in footnote 10 on page 92, sands should not be tested in an oven-dry state. Also moist soils should not be used because capillary forces may give the soil apparent cohesion. Research has shown, however, that there is a negligible difference between the strength properties of air-dry sand and saturated sand, provided the pore water is allowed to escape from the saturated sand during shear. The apparatus in Fig. X–8 can easily be adapted to testing saturated soils merely by placing the box in a pan of water.

Calculations

The angle of internal friction, ϕ, can be computed from

$$\phi = \tan^{-1}\left(\frac{\tau}{\sigma}\right) \qquad (X\text{-}1)$$

in which τ [15] = shear stress = (proving ring reading −
initial proving ring reading) × ring
calibration factor [16] ÷ area,

σ = normal stress = normal force ÷ area.

However, since τ/σ is a ratio of stresses on the same
surface, it can be determined by dividing the shear
force by the normal force. To compute the maximum
or peak friction angle, ϕ_m, use the maximum ratio,
$(\tau/\sigma)_m$ in Eq. X–1; and, to compute the ultimate
angle, ϕ_u, use $(\tau/\sigma)_u$.

Results

Method of Presentation. [17] The results of a direct
shear test on a cohesionless soil can be presented in
a summary table and/or by a stress-strain curve (or
stress ratio-strain curve). A table might give the
values of ϕ_m and ϕ_u with the shear displacement and
normal displacement at which each occurs (see page
95). A stress-strain curve usually consists of a plot
of τ/σ and normal displacement versus shear displacement (see Fig. X–12). The two plots should be on
the same page, using the same scale of shear displacement.

Typical Values. Peak friction angles in a dense,
well-graded, coarse sand usually range from 37° to
60°; for a dense, uniform, fine sand they are usually
between 33° and 45°. There is less variation in the
values of ultimate friction angle; a typical value is
30°. The volume changes that occur during shear
depend too much on the initial void ratio to give any
typical values. Normally, only the loosest of sands
tend to show a net negative volume change, or con-
traction, at failure.

---

[15] The method of computing the force depends, of course, on
the type of loading device used in the test. For example, the
force may be obtained by applying known weights, as is often
done in stress-controlled machines.

[16] For precise work, the calibration curve of the proving ring
should be used rather than the calibration factor. The factor
is the slope of the straight line which best fits the stress-strain
curve of the ring. For example, the calibration factor of the
ring used in the example on page 96 is 5 lb per 0.0001 in. of
ring deflection. Since the ring dial is an extensometer which
records the deflection of the ring, the dial reading minus the
initial reading multiplied by the factor is the force applied to
the ring. See pages 148–151, Appendix A, for a discussion of
proving rings and their calibration.

[17] Sometimes the results of a series of shear tests on sand are
plotted against pressure or depth similar to results of clay shear
tests. For a measure of in situ strength, a test is run in which
the applied normal stress is made equal to the estimated intergranular pressure to which the soil is subjected in nature.

Discussion. Under Discussion of Procedure were
presented some of the minor variables upon which
shear strength depends. Most of these are test details.
There are, however, two major variables which must
be considered.

The first variable is the denseness of the soil, usually
expressed as void ratio. As stated in the introduction
to this chapter, the denser the soil, the more the interlocking and, therefore, the greater the value of peak

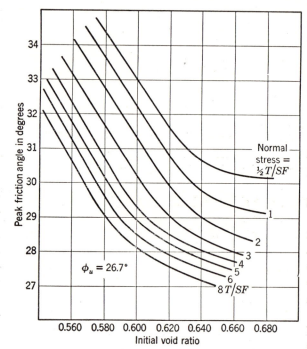

FIGURE X–11. Effect of void ratio on friction angle. (From
reference X–7.)

strength. Figure X–11 illustrates the magnitude of
the effect of denseness, as indicated by initial void
ratio, on the peak friction angle for Ottawa sand. The
ultimate friction angle is not much affected, however,
by the initial denseness. This seems reasonable since
a loose specimen and a dense specimen have approximately equal void ratios at ultimate failure, the dense
one having expanded and the loose one having con-
tracted during shear.

The second variable is the magnitude of applied
normal load (see page 105). Equation X–1 shows that
the shear resistance depends, of course, on the normal
load. The peak friction angle, ϕ_m, however, is also
dependent on the normal load; usually, the lower the
normal load, the higher ϕ_m. This relationship is illustrated in Fig. X–11. The ultimate friction angle is
not much affected by the normal load.

To a lesser degree, grain size and shape, grain mineral, and the distribution of grain sizes affect the fric-

tion angle. Smoothness, rounded corners, and uniform size of the soil grains tend to give lower friction angles. It is difficult to derive definite relationships between the friction angle and uniformity or grain size, because if either uniformity or grain size is varied, the void ratio of the soil obtained from a given amount of compaction is altered. The larger the maximum grain size and the less uniform the soil, the more dense the specimen which is obtained from a given compactive energy. Thus, in a series of tests, it is not easy to determine what percentage of the increased ϕ_m is due to lowered void ratio and what percentage is due to the better gradation of the soil or the larger maximum grain size (X–5).

Numerical Example

The soil used in the example on pages 96 and 97 was a uniform, fine sand in a loose state. The close agreement, therefore, between the peak and ultimate shear strengths would be expected. The latter portion of the normal displacement curve (Fig. X–12) is broken because leakage of sand from the shear box vitiated the displacement readings. This type of leakage, one of the disadvantages of the direct shear test, is discussed on page 99.

The results of the example are summarized below:

Peak angle, ϕ_m	$36\frac{1}{4}°$
Shear displacement at peak	0.180 in.
Normal displacement at peak	−0.0012 in.

Ultimate angle, ϕ_u	$35\frac{1}{2}°$
Shear displacement at ultimate	0.340 in.
Normal displacement at ultimate	−0.0005 in.

REFERENCES

1. American Society for Testing Materials, "Procedures for Testing Soils," July, 1950.
2. Casagrande, A., "Characteristics of Cohesionless Soils Affecting the Stability of Slopes and Earth Fills," *Journal of the Boston Society of Civil Engineering,* Vol. 23, No. 1, pp. 13–32, January, 1936.
3. *Engineering News-Record,* "Machine Speeds Soil Shear Tests," Jan. 26, 1950.
4. Housel, William S., "Laboratory Manual of Soil Testing Procedures," University of Michigan, 1950.
5. Leps, T. M., "The Effect of Gradation on the Shearing Properties of a Cohesionless Soil," Master of Science Thesis, Department of Civil and Sanitary Engineering, Massachusetts Institute of Technology, 1939.
6. Taylor, D. W., *Fundamentals of Soil Mechanics,* John Wiley and Sons, New York, 1948.
7. Taylor, D. W., and T. M. Leps, "Shearing Properties of Ottawa Standard Sand as Determined by the M.I.T. Strain-Control Direct Shearing Machine," *Record of Proceedings of Conference on Soils and Foundations,* Corps of Engineers, U.S.A., Boston, Mass., June, 1938.
8. Tschebotarioff, G. P., and J. D. Welch, "Lateral Earth Pressures and Friction between Soil Minerals," Paper V b 16, *Proceedings of the Second International Conference on Soil Mechanics,* Vol. VII, p. 135, 1948.
9. Waterways Experiment Station, "Triaxial Shear Research and Pressure Distribution Studies on Soils," *Progress Report,* Soil Mechanics Fact Finding Survey, Vicksburg, Miss., April, 1947.

SOIL MECHANICS LABORATORY

DIRECT SHEAR ON COHESIONLESS SOIL

SOIL SAMPLE _Sand; brown; fine, uniform subrounded particles, mostly quartz._

LOCATION _Union Falls, Maine_

BORING NO. _CA_   SAMPLE DEPTH _3.0 ft._

SAMPLE NO. _CA-1-3.0_

SPECIFIC GRAVITY, G_s, _2.67_

SOIL SPECIMEN WEIGHT

INITIAL WT. CONTAINER + DRY SOIL IN g _410.7_

FINAL WT. CONTAINER + DRY SOIL IN g _297.9_

WT. DRY SOIL USED, W_s, IN g _112.8_

TEST NO. _D 25_

DATE _Sept. 18, 1950_

TESTED BY _WCS_

SHEAR BOX

LENGTH, L, IN cm _7.62_

INSIDE DEPTH, b, IN cm _3.78_

BLOCK + GRATING, X, IN cm _4.83_

PROVING RING NO. _15 W_

CALIBRATION FACTOR _5 lb per 0.0001 in._

SOIL SPECIMEN VOLUME

BOX TO SOIL, y_a, IN cm _2.32_

SOIL THICKNESS, t, IN cm _1.27_

SAMPLE VOLUME IN cc _73.8_

SOLIDS VOLUME, V_s, IN cc _42.3_

VOID RATIO, e _.743_

POROSITY, n _.427_

NORMAL LOAD

APPLIED LOAD _750_ lbs. _12,000_ lbs./sq. ft.

TARE IN lbs. _45_

SCALE LOAD IN lbs. _795_

ELAPSED TIME IN min.	SHEAR DIAL IN in.	SHEAR DISPLACEMENT IN in.	NORMAL DIAL IN in.	NORMAL DISPLACEMENT IN in.	PROVING RING DIAL IN .0001 in.	SHEAR FORCE IN lbs.	$\frac{\tau}{\sigma}$
0	0.100	0.0	0.1567	−0.0	0.0	0.0	0.0
¼	.102	.002	.1561	−.0006	19.0	95.0	.126
½	.109	.009	.1547	−.0020	42.8	214.0	.286
¾	.119	.019	.1535	−.0032	61.0	305.0	.407
1	.129	.029	.1533	−.0034	71.5	357.5	.476
1¼	.140	.040	.1521	−.0046	81.5	407.5	.544
1½	.155	.055	.1520	−.0047	90.5	452.5	.603
1¾	.173	.073	.1521	−.0046	97.8	489.0	.651
2	.190	.090	.1525	−.0042	102.5	512.5	.683
	.220	.120	.1541	−.0026	107.2	536.0	.715
	.240	.140	.1546	−.0021	108.4	542.0	.724
	.260	.160	.1551	−.0016	109.3	546.5	.730
	.280	.180	.1555	−.0012	110.0	550.0	.733
	.300	.200	.1559	−.0008	110.0	550.0	.733
	.320	.220	.1561	−.0006	109.8	549.0	.732
	.340	.240	.1562	−.0005	109.0	545.0	.726
	.360	.260	.1562	−.0005	109.0	545.0	.726
	.380	.280	.1562	−.0005	108.9	544.5	.725
	.400	.300*	.1562	−.0005	108.2	541.0	.720
	.420	.320	.1562	−.0005	107.8	539.0	.718
	.440	.340	.1562	−.0005	107.1	535.5	.713
	.460	.360	.1561	−.0006	107.2	536.0	.715
	.480	.380	.1560	−.0007	107.5	537.5	.717
	.500	.400	.1558	−.0009	108.0	540.0	.720
	.520	.420	.1557	−.0010	108.0	540.0	.720
	.540	.440	.1555	−.0012	108.2	541.0	.722
6½	.560	.460	.1555	−.0012	108.0	540.0	.720

REMARKS:* _Sand observed leaking from shear box._

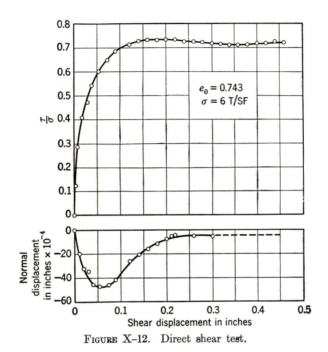

FIGURE X–12. Direct shear test.

CHAPTER

XI

Triaxial Compression Test on Cohesionless Soil

Introduction [1]

In the preceding chapter, the theory of shear strength in cohesionless soil was discussed, and three types of shear tests were described. It was stated that the two most widely used tests are the direct shear and the triaxial,[2,3] and that the triaxial is gaining in popularity. The reason for the trend can be understood from the following comparison of the direct shear and triaxial tests for testing cohesionless soils. In Chapter XIV, the comparison is extended to cover shear testing of cohesive soils.

The advantages of the triaxial test over the direct shear test are:

1. Progressive effects are smaller in the triaxial.
2. The measurement of specimen volume changes are more accurate in the triaxial.

[1] The student should read the introduction to this chapter prior to performing the triaxial test, and then reread it after completing the test.

[2] The word "triaxial" is commonly used for the name of this test, although it is not rigorously a test in which there is stress control on three axes. The chamber fluid makes the pressure on two principal planes planes always the same. A more proper name for the test is "cylindrical compression test." The two names will be used interchangeably in this book.

[3] Knowledge of triaxial testing has been greatly advanced by two extensive research programs sponsored by the Waterways Experiment Station of the Corps of Engineers. These programs were carried out at Harvard under A. Casagrande and at M.I.T. under D. W. Taylor. The results of these programs up to May, 1944, together with research done at the Waterways Experiment Station, have been summarized and reviewed by Rutledge (XI–13). Rather than refer to the M.I.T. and Harvard progress reports, we shall refer to the review (XI–13) wherever possible, since the progress reports are not generally available.

3. The complete state of stress is known at all stages during the triaxial test, whereas only the stresses at failure are known in the direct shear test.

4. The triaxial machine is more adaptable to special requirements.

The advantages of the direct shear test are:

1. The direct shear machine is simpler and faster to operate.
2. A thinner soil sample is used in the direct shear test, thus facilitating drainage of the pore water from a saturated specimen.

These points are discussed below.

Advantages of Triaxial

Progressive Effects. If a specimen is failed under a pressure system which gives a nonuniform distribution of stress on the failure surface, the entire strength of the specimen is not mobilized simultaneously for resistance to failure. Instead, the specimen is failed progressively, like the tearing of a piece of paper. Because progressive failure is fostered by nonuniform stresses and strains, it follows that the test which imposes the most uniform conditions on the test specimen will have the least progressive action.

Figure XI–1a shows a direct shear specimen before the start of shear. If the strains were uniform, the sheared specimen would be as shown in Fig. XI–1b; the vertical lines remain parallel but move to the positions shown. The strains of Fig. XI–1b, however, are not the actual ones, but those of Figs. XI–1c and XI–1d are.[4] In fact, the shear zone of a direct shear

[4] For a complete discussion and scaled drawings of displacement distributions in direct shear specimens, see reference XI–6.

specimen seems to be contained within the dotted lines of Fig. XI–1e. The soil at the edges of the box is usually failed before that at the center is close to failure.

Although the distribution of strains, and therefore stresses, is more nearly uniform in the triaxial test, it is not completely uniform. Figure XI–2 shows the cross section of a loaded triaxial specimen. Because

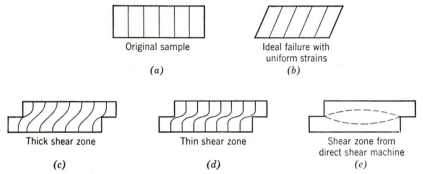

Original sample

(a)

Ideal failure with uniform strains

(b)

Thick shear zone

(c)

Thin shear zone

(d)

Shear zone from direct shear machine

(e)

FIGURE XI–1. Displacements in direct shear.

of the restraint furnished by the sample caps, there are dead zones at the top and the bottom in which practically no strains occur. The center zone, therefore, undergoes considerably greater strains than the edges; this is illustrated by the scales on the specimen, which were alike before shear. As a result of this strain distribution, the stresses in the center of the specimen are also considerably greater than those at the edge.

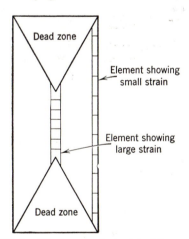

Dead zone

Element showing small strain

Element showing large strain

Dead zone

FIGURE XI–2. Displacements in triaxial compression.

From the point of view of stress distribution, the triaxial test is the superior of the two, because, although it does not give a uniform distribution, it gives one which is more nearly uniform than the direct shear test gives.

Volume Change Measurement. As pointed out in Chapter X, the volume changes which occur in a soil during shear can be extremely important. An indica-

tion of volume change is obtained in the direct shear test by measuring changes in the specimen thickness. This measurement is only an approximation of volume changes because the area of the shear surface varies and because soil usually leaks from the edges of the box during shear (see Fig. X–12). Since a saturated specimen is used in the triaxial test, precise measurements of volume change can be made by observing the volume of water which flows into or out of the specimen during shear.

State of Stress.[5] In the direct shear test, the complete state of stress on all planes within the soil can be statically determined only at failure. During testing, the normal and shear stresses on the horizontal plane are known; but in order for the stress system to be statically determinate, another condition must be known. The fact that the failure plane is horizontal furnishes the needed condition for the stress state to be determinate at failure. On the other hand, the boundary stresses in the triaxial test consist of known normal pressures and zero shear stresses on the horizontal and vertical planes. Therefore, the complete stress system is always statically determinate in the triaxial test.

Another way of expressing what has been stated in the preceding paragraph is to point out that a Mohr diagram can be drawn for any stage of the triaxial test,[6] whereas one can be drawn only for the failure condition in the direct shear test. In the triaxial test, the changes of stress on any plane can be traced as the test progresses; in the direct shear the stresses on the horizontal plane are the only ones known, except

[5] It is assumed that the reader is familiar enough with mechanics to be able to compute the stresses at a point, given the necessary information. In soils work, much use is made of the Mohr diagram to represent graphically the stresses on various planes in loaded element. For a discussion of the Mohr circle, the reader is referred to (XI–7).

[6] This is true of triaxial tests in which the pore-water pressures are known, such as the normal triaxial test on cohesionless soils the procedure of which is given in this chapter.

at failure. The fact that stress systems can always be determined is useful in research and, as pointed out below, will become important for practical testing as soon as more is learned about the stress systems in soils in nature.

Adaptability of Test. By its nature, triaxial apparatus is considerably more adaptable to special requirements. The triaxial specimen can be failed in tension or compression, drainage can be completely prevented, the boundary stresses can be altered during testing, etc. The triaxial test, therefore, is more valuable for research purposes and for other testing in which control of the boundary conditions is wanted.

Advantages of Direct Shear

Simplicity of Operation. Since the direct shear machine is simpler to operate than the triaxial equipment, a less skilled technician is required for direct shear testing. Also less testing time is required in the direct shear test; a skilled technician can perform two to three direct shear tests in the time that he requires to run one triaxial test.

Drainage Facilities. Since a thin sample is employed in the direct shear test, the time required for any pore water to drain out of the sample is small. This point is less important in sand testing than in clay testing because of the relatively large permeability of sands. More discussion is given to this point on page 138.

The purpose of any laboratory shear test is to secure information on the effective strength of the soil in the field. Remembering this fact, we can reason that the best type of shear test would duplicate field conditions. Unfortunately, little is known of the pressure systems which exist in soils in situ, although much study of this major problem is going on. A discussion of this work is not within the scope of a book on laboratory testing.

Since the pressure systems existing in soils in nature are not completely known, the best type of laboratory shear test is the one in which the conditions of stress and strain are understood and controlled. Such a test will, of course, increase in value as knowledge of stresses and strains in the ground increases. On this reasoning is based the preference of the triaxial testing over direct shear testing. To prove that either test more nearly duplicates conditions in nature would be a difficult problem; to prove, however, that the conditions are better understood and more controllable in the triaxial test would be easy.

From the preceding comparison, it can be seen why the triaxial test [7] is gaining favor at the expense of the

direct shear test. Nevertheless, the speed and simplicity of the direct shear testing justify its continued use for testing cohesionless soils.

Apparatus and Supplies

Special

1. Triaxial machine
2. Specimen dowel
3. Specimen mold

General

1. Tamper for compacting sand
2. Rubber [8] membrane
3. Small level (bull's-eye level)
4. Deaired water supply
5. Vacuum supply
6. Balance (0.1 g sensitivity)
7. Drying oven
8. Length gage
9. Large evaporating dish
10. Rubber strips for binding
11. Spoon
12. Timer

The triaxial machine, as the direct shear, can be stress-controlled or strain-controlled. The relative merits of the two methods of shear given on page 91 hold for the triaxial machine; there is an additional drawback, however, to stress control. The setup for a stress-controlled machine normally requires that the magnitude of the applied axial load be measured outside the chamber. Therefore, any friction between the piston which applies the axial load and the rest of the apparatus will constitute an unknown error. This problem has received considerable study (XI–12) and can be overcome to a large extent by careful design of apparatus details.

For shear and consolidation tests, methods of applying and measuring loads are necessary. Among the many methods used are: [9]

1. Wheel or lever systems employing known weights (see Fig. IX–4).

2. Jack units for applying loads which are measured by platform scales (see Figs. IX–3 and XII–2).

3. Proving rings [10] (see Figs. X–8, XII–3, and XII–4).

4. Hydraulic and air pressure measured by gages (lateral pressure in Fig. XI–3).

5. Application of weights which are measured by balances (reference XI–12).

---

[7] The triaxial test is gaining in acceptance for use with soil-bituminous mixtures in pavement design.

[8] There has been some use of membranes made of synthetic plastics.

[9] Hough of Cornell has suggested one loading machine which can be easily adapted for several tests.

[10] Also proving frames are used. See Fig. A–4.

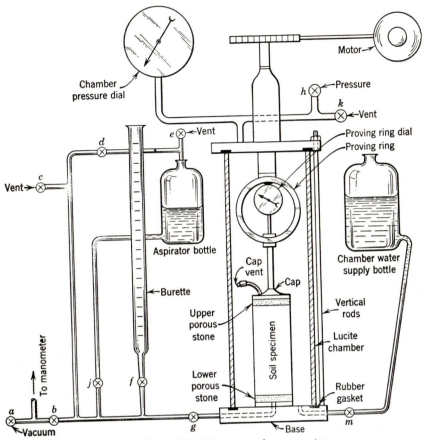

FIGURE XI–3. Triaxial compression apparatus.

FIGURE XI–4. Accessories for triaxial test.

In parentheses following methods 1 through 4 are listed figures which show devices using that particular method. Method 5 is normally applicable for small loads. Proving rings are the best method of load measurement for strain-controlled shear apparatus. On page 148, Appendix A, is a discussion of proving rings which includes the procedure of calibrating the rings. The choice of loading method depends on the type of apparatus, the type and size of laboratory, and the personal preference of the designer.

Figure XI–3 is a diagrammatic sketch of a strain-controlled triaxial machine. This sketch will be used for the test procedure which is presented in the following pages. Figure XI–4 shows some accessories used in the triaxial test. On the left of Fig. XI–4 is a spool of rubber strip; in the center is a dowel with a membrane slid partly over it; at the sides of the dowel are the halves of a specimen mold; and in the foreground is a level. In Fig. XI–5 more of the accessories can be seen.

Recommended Procedure [11]

1. Obtain the thickness of the membrane. This thickness is best obtained by measuring the membrane doubled and then halving the measurement.

2. Roll up the membrane and slide it on the dowel with about ½ in. of it projecting from one end of the dowel.

3. Moisten this projecting end and place it over the base (see Fig. XI–3) of the apparatus which contains the lower porous stone.[12]

4. Bind [13] the membrane to the base with a rubber strip and remove the dowel.

5. Clamp the mold around the membrane and turn the top end of the membrane over the top of the mold.

6. Weigh to 0.1 g a dish with the dry [14] soil [15] which is to be tested.

7. Place the sand within the membrane by tamping each spoonful of soil, taking care not to pinch the membrane with the tamper (see Fig. XI–5). Scarify the top of each layer before placing the next one, to

[11] This test can be done better by two or more students. They should be able to perform their first test in about 3 hours and do the computations in 2 to 3 hours. They will need supervision for most of the test.

[12] Stone, carbon, and porous brass (or bronze) are used for drainage plates. Carbon and brass can be more easily shaped than stone.

[13] See footnote 6 on page 124.

[14] See footnote 10, page 92.

[15] It is desirable to keep the diameter of the largest soil particle a small fraction of the diameter of the test specimen, e.g., one-tenth to one-fifteenth.

reduce stratification. The amount of tamping depends on the denseness of soil desired.

8. Again weigh the dish of soil. The difference in weights is the weight of soil used.

9. Put the upper porous stone and cap on top of the specimen and level by means of the level bubble.

10. Moisten the upper end of the membrane, roll up over the sides of the cap, screw in one vertical rod.

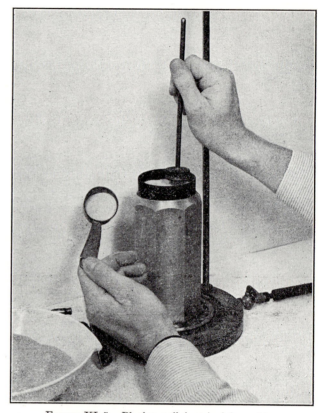

FIGURE XI–5. Placing soil in triaxial machine.

attach a clamp from the cap to the rod, thus lining up and supporting the cap, and carefully bind the membrane to the cap with a rubber strip.

11. Close all valves (*a* to *m*, inclusive) except *k* and *e*, which are left open.

12. The specimen is now saturated by first evacuating it and then admitting deaired water. To evacuate, close the cap vent and apply a vacuum by opening *a*, *b*, and *g*. To saturate, close *a* and *b* and then open *j*.

13. After time has been allowed for saturating, open the cap vent and let the water flow out. This insures that water got all the way up the specimen height. Close the cap vent after this check is made.

14. Next a vacuum of 5 in. of mercury [16] is applied to the specimen by first opening valves *b*, *d*, *j*, and *g*

[16] This is an arbitrarily selected, small pressure.

while closing *e*, and then carefully admitting the vacuum by opening valve *a*.

15. With the specimen under this 5 in. of vacuum, remove the cap clamp, check the level of the cap, and then carefully remove the sample mold.

16. After the mold is removed, increase the vacuum to 10 in.[16] by opening valve *a* slightly.

17. Measure the length of the specimen to 0.1 mm and measure the circumference of the specimen at the top, midpoint, and bottom to 0.1 mm by means of a tape.

18. Remove the level and clamp, then screw the remaining vertical rods in the base.

19. Next wet the bottom rubber gasket, center the lucite cylinder [17] on this gasket, wet the upper rubber gasket, and place it on top of the lucite cylinder.

20. Carefully put the upper assembly of the machine in place, then check to see that the plunger contacts the sample cap at its center.

21. Tighten [18] all the top nuts on the vertical rods until they just begin to bind, and then give each one-fourth of a revolution turn. Keep giving each nut one-fourth turns until two complete turns have been given.

22. Check with a length gage to make sure that the upper plate is parallel to the base.

23. Admit water [19] to the chamber by opening valve *m*; bring the water level up until the cap is just covered, then close valve *m*.

24. Close valve *k*, open the pressure drum, and set the desired chamber pressure [20] on the drum regulator.

25. Carefully build up the chamber pressure to 5 psi by opening valve *h*; at the same time, release the sample vacuum by opening valve *c*. The pressure build-up and vacuum release should be so synchronized that at all times the sum of the pressure and vacuum is 10 in. of mercury [21] (or 5 psi); e.g., when the chamber pressure is 3 psi, the vacuum should be 4 in. of mercury.

26. After the vacuum has been completely released, close valves *b*, *d*, and *j*; leave valve *c* open and open valve *f*.

[17] If high chamber pressures are to be employed, the cylinder should be reinforced with circumferential bands. Safety glass shields between the cylinder and the technician are also suggested for high-pressure work.

[18] A torque wrench is often employed to control the tightening.

[19] Other chamber fluids are used. See page 129.

[20] Thirty psi is often used when no special pressure is desired.

[21] As pointed out in footnote 16, there is nothing fundamental about the 5 and 10 in. of mercury used. Step 25 is designed to keep the pressure on the specimen constant from the time at which it was measured (step 17) to the time the volume change measurements are started (step 28).

27. The water level in the burette should be brought to a reading about the midpoint of its scale in order that either expansions or contractions can be measured. (Water can be added to the burette during the test if necessary.)

28. Record the burette reading, slowly bring the chamber pressure to its full value, which is maintained constant throughout the test, and again record the burette reading. (The difference of these burette readings is the volume change the specimen has undergone because of the chamber pressure increase.)

29. At this point, check for any leakage through the membrane; a leak is indicated by a rise of the water level in the burette.

30. Lower the plunger until it is in contact with the specimen cap.

31. At this stage, carefully check to see that all is ready to begin testing. A student should check with his instructor.

32. Record initial readings of proving ring dial, motor revolution counter,[22] burette and time, and then start loading.

33. For the first 2% of axial strain, take a set of readings about every 0.2% of strain. For the rest of the test, take readings every 0.5% to 1% strain. Time observations need only be made every third or fourth set of regular readings.

34. Continue the test until the compressive force remains constant for a few readings or until the specimen has been compressed approximately 15%.

35. Remove the axial force and again check for membrane leakage.

36. Read the burette, close valve *h*, release the chamber pressure through valve *k* until it is 5 psi, and then reread the burette. The difference in burette readings is the change in specimen volume caused by the decrease in lateral pressure.

37. Close valves *f* and *c*, open valves *b*, *j*, and *d*; apply a vacuum of 10 in. of mercury through valve *a* and release the chamber pressure to zero through valve *k*. This release of pressure and application of vacuum should be so synchronized that the equivalent of 10 in. of mercury vacuum is always on the soil specimen.

38. Drain the chamber water by lowering the supply bottle and by opening valve *m*.

39. Disassemble the apparatus (removing the chamber nuts one-half to one revolution at a time).

40. Sketch the failed specimen. On the sketch, dimension the maximum and minimum diameters, the

[22] The count of motor revolutions is used to compute axial compression. See Eq. XI–1.

length of the specimen, and the angle of inclination of the failure plane, if there is one.

41. Release the vacuum and remove the specimen.

Discussion of Procedure

Type of Tests. The procedure presented above is one in which the lateral pressure is kept constant. For general sand testing, this method is the easiest to perform and to interpret; therefore, it enjoys the widest use. Nevertheless, occasions arise in which some other method is desirable. For example, the rapid shear of a large mass of soil is more nearly duplicated by a triaxial test in which the volume of the soil specimen is kept constant by altering the lateral pressure (constant volume test). Other tests include one in which the volume is kept constant by altering the pore water pressure (closed system test); one in which the axial pressure is kept constant and the lateral pressure varied; and several types of axial extension tests.

Triaxial tests on dry cohesionless soils are run in which the lateral pressure is obtained by applying a vacuum to the soil. The maximum lateral pressure that can be obtained is 1 atmosphere or 14.7 psi.

In Europe, particularly in Holland and in Belgium, a test known as the cell test is used. This test, somewhat similar to the triaxial test, fails a cylindrical specimen of soil by increasing the vertical stress and measuring the resulting lateral stress. Although many cell tests are run, the significance of the results is nebulous.

Specimen Size and Shape. Research (XI–13) has indicated that the shape of the cross section of a specimen does not affect the test results. Although studies on the effect of specimen size are not complete enough for us to draw definite conclusions, they do suggest that higher unit strengths are obtained on smaller test specimens. There is evidence (XI–13) that ratios of length to diameter of 1.5 to 3.0 are satisfactory; a ratio of $2\frac{1}{2}$ is commonly employed. Below a ratio of 1.5, the test results are seriously affected, as would be expected from a study of Fig. XI–2. Two specimen sizes which are widely used are 1.4 in. in diameter and approximately 3.5 in. in height; 2.8 in. in diameter and approximately 6.5 in. in height.[23]

Employing a segmented cap which did not offer lateral restraint to the specimen, Taylor (XI–13) found that the dead zones shown in Fig. XI–2 have negligible effect on the test results. He used speci-

mens which were 2.8 in. in diameter and approximately 6.5 in. in height.

Rate of Strain. As with direct shear, the rate of strain in triaxial sand testing does not appear to be important as long as the pore water is permitted to flow freely in or out of the soil to prevent the build-up of excess pressures. Tests on one type of dry sand indicated only a 10% increase in strength when the elapsed time from the start of loading to the time of maximum compressive stress was decreased from 1000 seconds to 0.01 second (XI–2). Rates of strain in the neighborhood of $\frac{1}{4}\%$ to 2% per minute are satisfactory for routine testing.

Degree of Specimen Saturation. The procedure of admitting water to the specimen recommended in step 12 results in a degree of saturation of essentially 100%, whereas the free flow of water with no evacuation gives approximately 80% saturation. Since air in a partially saturated sand might influence volume change measurements and add to the apparent strength of the soil by introducing capillary pressures, the slight additional trouble of obtaining saturation is justified.

Membrane Influence. A. Casagrande (XI–13) has shown that even very thin membranes influence the test results by giving the soil specimen added stiffness. Chen (XI–13) found that the membrane had important effects on the slope and shape of the stress-strain curve, but little influence on the maximum stress.

Calculations

The change of specimen length, ΔL, is found from equation [24]

$$\Delta L = a - b \qquad (XI–1)$$

where a = movement of the top of the proving ring = k (in inches per revolution) × number of revolutions. k is a property of the gear box, and is found by calibration.

b = compression of proving ring = proving ring dial reading in inches − initial proving ring dial reading in inches.

The axial strain, ϵ, at any time is found from

$$\epsilon = \frac{\Delta L}{L_0} \qquad (XI–2)$$

where L_0 = the initial specimen length.

---

[23] A specimen 2.8 in. in diameter has been recommended as the minimum size (XI–13).

[24] The change of specimen length is more easily measured in the stress-controlled apparatus by an extensometer which is connected to the piston. Although not convenient, such a direct method can be used on the strain-controlled machine, by placing the extensometer within the chamber. See Fig. XIII–7.

The average cross-sectional area, A, at any time is found from equation [25]

$$A = \frac{A_0}{1 - \epsilon} \quad (XI-3)$$

where A_0 = the initial cross-sectional area of specimen.

The applied stress, p, is found from

$$p = \frac{P}{A} \quad (XI-4)$$

where P = applied force = (proving ring reading − initial proving reading) × ring calibration factor.[26]

The principal intergranular stress ratio, $\bar{\sigma}_1/\bar{\sigma}_3$, is found from

$$\frac{\bar{\sigma}_1}{\bar{\sigma}_3} = \frac{p + \bar{\sigma}_3}{\bar{\sigma}_3} \quad (XI-5)$$

where $\bar{\sigma}_1$ = the major principal intergranular stress = $p + \bar{\sigma}_3$,

$\bar{\sigma}_3$ = the minor principal intergranular stress = the chamber pressure for a test where the water pressure is zero, such as the one for which the procedure was given in the preceding pages.

The friction angle, ϕ, is computed from

$$\phi = \sin^{-1} \frac{(\bar{\sigma}_1/\bar{\sigma}_3) - 1}{(\bar{\sigma}_1/\bar{\sigma}_3) + 1} \quad (XI-6)$$

If ϕ_m is desired, use $(\bar{\sigma}_1/\bar{\sigma}_3)_m$ in Eq. XI–6; and if ϕ_u is desired, use $(\bar{\sigma}_1/\bar{\sigma}_3)_u$ in Eq. XI–6. In a test in which the stress-strain curve does not become horizontal, thus not indicating an ultimate strength, the strength at 15% strain is often designated the ultimate strength.

The angle, θ, between the horizontal (actually the major principal plane) and the failure plane at the time of its formation can be computed from

$$\theta = 45° + \frac{\phi_m}{2} \quad (XI-7)$$

The slope of the initial portion of the stress-strain curve is often computed; this slope is called the modulus of deformation.

Results

Method of Presentation.[27] The results of a triaxial test on cohesionless soil can be presented in a summary table and/or by a stress-strain curve (or stress-ratio versus strain curve). A table might give the values of ϕ_m and ϕ_u, with the strain and volume change at which each occurs. Also the void ratio existing when the specimen was measured and that existing when the chamber pressure was applied should be included (see page 106). A stress-strain curve usually consists of a plot $\bar{\sigma}_1/\bar{\sigma}_3$; normally a plot of volume change versus strain is also given. (See Fig. XI–7.) Both plots should be on the same page and employ the same strain scale.

Typical Values. Typical values of peak and ultimate friction angles for cohesionless soils were given on page 94.

Discussion. On page 89, the critical void ratio of a soil was discussed. The most widely used one, known as the Casagrande critical void ratio, can be determined by using the test procedure described in this chapter. From a series of tests, a plot of initial void ratio against volume change at failure is made. From such a plot, we can obtain the critical void ratio by taking that initial void ratio which showed no volume change at failure. Casagrande used the void ratio before the application of lateral pressure as the initial one, and the peak strength as the failure point.

It was pointed out on page 94 that the friction angle depends on the details of the test procedure used, void ratio, and magnitude of applied stress. Chen (XI–3, XI–13), found that ϕ_m is almost constant for a given soil tested by the same procedure at the same void ratio but with varying lateral pressures. He also showed that the plot of shear stress against normal stress [28] had a shear strength intercept of C_1 at zero normal stress (see Fig. XI–6). C_1 appears to be smallest when determined by tests in which the experimental errors are at a minimum. Rutledge (XI–13) suggests that C_1 may be caused by some sort of surface effect on the specimen. The angle, ϕ, determined from Eq. XI–6, has been named the "apparent friction angle"[29] and ϕ' the "corercted friction angle" (XI–13). The corrected angle is smaller than the apparent angle. Both angles are shown in Fig. XI–6.

plotted against pressure or depth similar to results of clay shear tests (see page 118). For a measure of in situ strength, a test is run in which $\bar{\sigma}_3$ is made equal to the estimated intergranular pressure to which the soil is subjected in its in situ condition.

[28] Plots of normal against shear stress for the triaxial test are discussed in Chapter XIII. A Mohr circle for a triaxial test on cohesionless soil is drawn similar to that for a drained triaxial test on a clay. See Fig. XIII–9.

[29] Note that this apparent friction angle is not the same one used in describing the shear strength of clay.

[25] A plot of ΔL against A can be made for any given L_0 and A_0; for example, see Figs. XII–7 and XIII–8. Equation XI–3 is only approximate because the conditions on which its derivation is based are not met. See page 160. Equation XIII–5, page 129, can be used in place of Eq. XI–3, if allowance for volume changes is desired.

[26] See footnote 16 on page 94.

[27] Sometimes the results of a series of shear tests on sand are

Taylor (XI–5, XI–11) has suggested that a cohesionless soil has a friction angle made up of two components—true friction and a portion due to volume changes. He pictures the true friction angle as the ultimate angle.

Moderately good agreement of results has been obtained from direct shear and triaxial compression tests, although the tests are entirely different. We

Numerical Example

In the example on pages 108 and 109 the computed angle of 66° between the shear plane and the horizontal does not agree very well with the observed value of 56°. This discrepancy can be partially explained by the fact that the angle θ was measured at the end of the test rather than at the time the failure plane was formed.

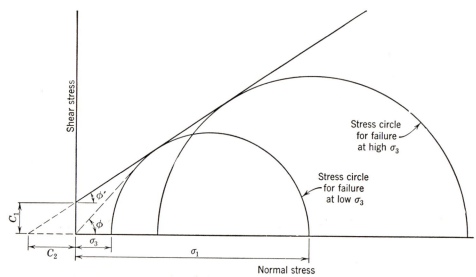

FIGURE XI–6. Mohr circles for triaxial compression on cohesionless soil.

should expect higher strength from triaxial because of smaller progressive action; on the other hand, we should expect lower strength [30] because the intermediate principal is merely equal to the minor principal stress in this test (i.e., triaxial test), whereas it is greater than the minor principal stress in the direct shear test. Critical void ratios determined by the direct shear test are measurably larger than those obtained from the triaxial test (XI–10).

A triaxial or unconfined (see Chapter XII) soil specimen may fail in one of two general ways—a plastic flow failure or a shear failure. In a plastic flow failure, the specimen bulges to a barrel shape without the formation of a definite rupture surface, whereas a distinct rupture surface is developed in a shear failure. Both failures are illustrated in Fig. XII–9. The failure of the undisturbed specimen is a shear one and that of the remolded is a plastic flow one.

[30] A lowered intermediate principal stress does not appear to cause a decrease in the strength of a cohesive soil. Clough of M.I.T. obtained 15% to 20% greater strength when $\bar{\sigma}_2 = \bar{\sigma}_3$ than when $\bar{\sigma}_2 = \sigma_1$ for undrained triaxial shear on undisturbed Boston blue clay.

Figure XI–7 shows a volume increase at peak strength; therefore, the specimen tested was at an initial void ratio less than its critical one. Figure XI–7 is a typical plot for a dense, well-graded, coarse sand; Fig. X–12 represents the other extreme, a loose, uniform, fine sand.

The results of the following example are summarized here:

Peak angle, ϕ_m	41.9°
Strain at peak	4.0%
Volume change at peak	+10 cc
Ultimate angle, ϕ_u	35.6°
Strain at ultimate	15%
Volume change at ultimate	+36 cc
Initial void ratio at 5 psi	0.453
Initial void ratio at 30 psi	0.452

REFERENCES

1. Casagrande, A., "Reports on Co-operative Research on Stress-Deformation and Strength Characteristics of Soils," submitted to Waterways Experiment Station, Harvard University, 1940–1944. Unpublished, reviewed in Reference 13.
2. Casagrande, A., and W. L. Shannon, "Stress–Deformation and Strength Characteristics of Soils under Dynamic Loads," *Proceedings of the Second International Conference on Soil Mechanics*, Vol. 5, Paper II d 10, p. 29, 1948.

3. Chen, Liang-Sheng, "Stress-Deformation and Strength Characteristics of Cohesionless Soils," Doctor of Science Thesis, Harvard University, Dec., 1944.

4. Fidler, H. A., "Investigation of Stress-Strain Relationships of Granular Soils by a New Cylindrical Compression Apparatus," Doctor of Science Thesis, Department of Civil and Sanitary Engineering, Massachusetts Institute of Technology, January, 1940.

5. Hsiao, Feng, "An Investigation of the Effect of Rate of Volume Change on the True Friction Angle of Cohesionless Soils by Cylindrical Compression Tests," Master of Science Thesis, Massachusetts Institute of Technology, February, 1946.

6. Hvorslev, M. J., "Über die Festigkeitseigenschaften Gestorter Bindiger Boden," Danmarks Naturvidenskabelige Samfund, 1 Kommission Hos G.E.C. Gad, Vimmelskaftet 32, Kobenhavn, 1937.

7. Taylor, D. W., *Fundamentals of Soil Mechanics*, John Wiley and Sons, New York, 1948.

8 Taylor, D. W., "Shear Investigation of Granular Soils," Master of Science Thesis, Department of Civil and Sanitary Engineering, Massachusetts Institute of Technology, May, 1942.

9. Taylor, D. W., "Reports on Co-operative Research on Stress-Deformation and Strength Characteristics of Soils," submitted to Waterways Experiment Station, Massachusetts Institute of Technology, 1940–1944. Unpublished, reviewed in Reference 13.

10. Taylor, D. W., "A Comparison of Results of Direct Shear and Cylindrical Compression Tests," *Proceedings of the American Society for Testing Materials,* Vol. 39, 1939.

11. Tsien, Shou-I, "The Effect of Rate of Volume Change on the Friction Angle of Cohesionless Soil," Master of Science Thesis, Massachusetts Institute of Technology, February, 1944.

12. Warlam, Arpad A., "Stress-Strain and Strength Properties of Soils," Doctor of Science Thesis, Harvard University, May, 1946.

13. Waterways Experiment Station, "Triaxial Shear Research and Pressure Distribution Studies on Soils," *Progress Report,* Soil Mechanics Fact Finding Survey, Vicksburg, Miss., April, 1947.

SOIL MECHANICS LABORATORY

TRIAXIAL COMPRESSION TEST ON COHESIONLESS SOIL

SOIL SAMPLE *Sand: brownish gray; coarse well graded, subrounded particles; mostly quartz & feldspar, some mica.*
LOCATION *Union Falls, Maine*
BORING NO. *CD* SAMPLE DEPTH *4.0*
SAMPLE NO. *CD-1-4.0*
SPECIFIC GRAVITY, G_s, *2.67*

SOIL SPECIMEN WEIGHT
INITIAL WT. CONTAINER + DRY SOIL IN g *1644.5*
FINAL WT. CONTAINER + DRY SOIL IN g *531.0*
WT. DRY SOIL USED, W_s, IN g *1113.5*

TEST NO. *T 21*
DATE *Dec. 11, 1949*
TESTED BY *WCS*

SOIL SPECIMEN MEASUREMENTS
CIRCUMFERENCE TOP *22.50 cm*
MIDDLE *22.65*
BOTTOM *22.55*
AVERAGE *22.57*
MEMBRANE THICKNESS *0.023 in.*
NET CIRCUMFERENCE *22.20 cm*
INITIAL LENGTH, L_o *15.50 cm*

PROVING RING NO. *2*
CALIBRATION FACTOR *5 lb. per 0.0001 in.*

SOIL SPECIMEN VOLUME
INITIAL AREA, A_o, IN sq. cm. *39.15*
VOLUME, A_oL_o, IN cc *606*
SOLIDS VOLUME, V_s, IN cc *417*
INITIAL VOID RATIO, e_o,
AT *5* lbs./sq. in. *.453*
AT *30* lbs./sq. in. *.452*

VOLUME CHANGE, ΔV
BURETTE READING
INITIAL
AT *5* lbs./sq. in. *8.2*
AT *30* lbs./sq. in. *7.6*
ΔV *0.6*

FINAL
AT *30* lbs./sq. in. *44.6*
AT *5* lbs./sq. in. *44.7*
ΔV *0.1*

CHAMBER PRESSURE, σ_{ch}, IN lbs./sq. in.	COUNTER	PROVING RING DIAL IN .0001 in.	BURETTE IN cc	ELAPSED TIME IN min.	LENGTH CHANGE, Δl, IN in.	$\frac{\Delta L}{L_o} = \epsilon$ IN %	AREA, A, IN sq. in.	AXIAL LOAD, P, IN lbs.	P/A IN lbs./sq. in.	$\frac{\bar{\sigma}_1}{\bar{\sigma}_3}$	VOLUME CHANGE, ΔV, IN cc
30.0	22000	0.3	7.6	0	0.0	0	6.06	0.0	0.0	1.00	0.0
	10	14.0	7.3	½	.009	0.2	6.07	68.5	11.3	1.38	−0.3
	20	35.4	6.9	¾	.017	0.3	6.08	175.5	28.8	1.96	−0.7
	30	50.3	6.6		.026	0.4	6.09	250.0	41.0	2.36	−1.0
	40	65.8	6.6		.035	0.6	6.10	327.5	53.8	2.79	−1.0
	60	89.7	6.5	2½	.053	0.9	6.12	447.0	72.8	3.42	−1.1
	80	107.0	6.7	3¼	.072	1.1	6.13	533.5	87.1	3.91	−0.9
	22100	119.6	7.4		.092	1.5	6.16	596.5	96.8	4.23	−0.2
	20	129.9	8.3	4¾	.111	1.8	6.18	648.0	105.0	4.50	+0.7
	60	141.5	10.3	6½	.152	2.5	6.20	706.0	114.0	4.80	+3.2
	22200	148.5	13.4	8	.192	3.2	6.27	741.0	118.1	4.94	+5.8
	50	150.3	17.4	10¼	.244	4.0	6.32	750.0	118.8	4.96	+9.8
	22300	150.5	21.3	12¼	.296	4.9	6.37	751.0	118.0	4.94	+13.7
	50	148.6	24.9		.347	5.7	6.43	741.5	115.1	4.84	+17.3
	22400	148.6	28.2	16¼	.399	6.6	6.49	741.5	114.1	4.81	+20.6
	50	148.7	31.1		.451	7.4	6.55	742.0	113.2	4.78	+23.5
	22500	133.0	33.5	20¼	.504	8.3	6.61	663.5	100.4	4.35	+25.9
	50	131.0	35.4		.556	9.1	6.67	653.5	98.1	4.27	+27.8
	22600	127.0	37.2	24½	.608	10.0	6.74	633.5	94.0	4.13	+29.6
	50	124.9	38.4		.661	10.8	6.80	623.0	91.8	4.06	+30.8
	22700	124.7	39.7		.712	11.7	6.87	622.0	90.5	4.02	+32.1
	50	121.9	40.8		.764	12.5	6.94	608.0	87.6	3.92	+33.2
	22800	122.2	41.6	32½	.816	13.4	7.01	609.5	87.0	3.90	+34.0
	50	122.7	42.5	34½	.867	14.2	7.07	612.0	86.6	3.89	+34.9
	22900	120.9	43.3		.819	15.1	7.15	603.0	84.5	3.82	+35.7
	50	121.9	43.9	38¾	.971	15.9	7.21	608.0	84.3	3.82	+36.3
	23000	120.9	44.6	40¾	1.023	16.8	7.29	603.0	82.8	3.76	+37.0

REMARKS:

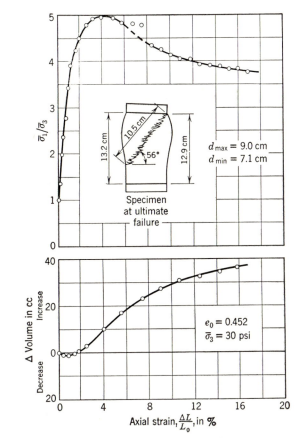

FIGURE XI–7. Triaxial compression test.

CHAPTER

XII

Unconfined Compression Test

Introduction

Strength Theory of Cohesive Soil. The two preceding chapters were concerned with shear testing of cohesionless soils; this chapter and the following two are devoted to shear testing of cohesive soils. The elements of shear strength are considerably more complicated in cohesive soils than in cohesionless soils because of the more complex make-up of cohesive soils. Cohesionless soils are composed of particles which, because of their size and shape, have a small specific surface; that is, ratio of surface area to mass. The mass forces, such as gravity, control their behavior rather than surface forces. Particles of cohesive soil, on the other hand, because of their size and shape (usually small plate shapes, see Fig. IV–5), have a large specific surface. Their behavior, therefore, can be influenced more by surface forces than by mass forces.

The shear strength of a cohesive soil is made up of two components: friction, as in cohesionless soil, and a second component called "cohesion," which is all the strength not due to friction. The exact nature of the surface forces which cause cohesion is not known. The cohesion of a soil is not a constant soil property but is a function of the load carried by the soil structure, or intergranular load, as well as of the details of the test by which it is determined. The term cohesion is often used loosely for the shear strength of a soil when tested with no lateral load applied to the specimen.

The following analogy will help clarify this. In Fig. XII–1b is plotted the shear force required to move the block shown in Fig. XII–1a for different applied normal forces. If the weight of the block is neglected, the relationship between normal force and

shear force required to move the block is shown by line OA. Line OA is analogous to a plot of normal force against shear strength for a cohesionless soil.

If the block in Fig. XII–1a is now stuck to the surface on which it rests with some adhesive which can

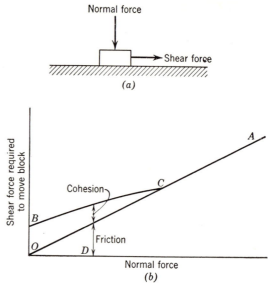

FIGURE XII–1. Analogy of cohesion.

be squeezed out by pressure, line BCA in Fig. XII–1b represents the relationship between an applied normal force and the shear force required to move the block. At any normal force D, the resistance to movement is due to friction between the block and surface (corresponding to friction between particles in soil) plus the strength of the adhesive (corresponding to cohesion in soil). At a normal force of C, all the adhesive is squeezed out of the space between the block and surface. In like manner, the cohesion of a soil can be reduced to zero at large normal pressures. The loose definition of cohesion mentioned above would be represented by *OB*, which is the shear force required to move the block when no normal force is applied.

The portion of a load applied to a cohesive soil which is carried by the soil structure depends on the degree to which the pore water is permitted to drain and thus release its hydrostatic excess pressures (i.e., consolidate, see Chapter IX). Since the load carried by water is not able to mobilize friction between soil particles, the shear strength of clay is higher [1] if drainage occurs than if it is prevented. The two limiting drainage possibilities are: (1) shear where there is complete pore water drainage making the excess pressure in the water zero, and (2) shear where no water drainage occurs. The first type of shear is named "drained" or "slow" [2] shear; the second type is called "undrained" or "quick" [2] shear.

Although both drained and undrained shear are possible in nature, most actual cases lie between these two limits. For an example of the limiting cases, the clay under a small footing may be loaded over a period of years so gradually that it shears slowly enough to insure complete drainage. On the other hand, the shear of an earth dam because of a rapid lowering of the pond may occur so fast that it prevents the clay core from draining at all. For each practical case, the engineer has to use his judgment in deciding whether a possible shear failure would occur with drainage, without it, or with partial drainage, the degree of which has to be estimated.

A complete understanding of shear strength in cohesive soils is difficult to obtain because the fundamentals of cohesion are not known at the present time. An undisturbed clay when sheared may, initially, have a large amount of cohesion, but only a small amount after it has been thoroughly kneaded, or remolded. This loss of strength through remolding, which is said to be caused by the fact that the soil possesses "structure," [3] is defined as "sensitivity," [3] and sensitivity is used as a measure of structure.

Shear Tests on Cohesive Soil. When using laboratory tests to determine the shear strength of a soil involved in a problem, the engineer has the choice of two approaches. He can test samples of the soil, each under a different pressure system, to determine the relationship between shear strength and pressure.[4] He can then prepare a plot of strength against pressure, known as a strength envelope, which is analogous to line *BCA* in Fig. XII–1*b*. After determining the pressure system to which the soil will be subjected in nature, he can predict the effective shear strength from his predetermined pressure-strength curve. The other approach is to determine the in situ shear strength of the soil, and use this for the effective strength. If the soil at a particular site consists of a number of different strata, the first approach becomes impractical because of the many tests required to obtain an envelope for each stratum. To determine the in situ strength of each stratum would be more economical. On the other hand, the pressure system within the soil may be altered by construction or the later additions of loads in such a way that the original in situ strength is not a good approximation of the effective strength.

Any of the methods of shearing described on pages 89 and 90 can be used in either of these approaches. Direct shear and triaxial shear methods for both approaches are presented in Chapters XIII and XIV. In this chapter the unconfined compression test is presented. It can be considered a special form of the triaxial test. The unconfined compression test measures the compressive strength of a cylinder of soil to which no lateral support is offered. The shear strength [5] is taken as equal to half the compressive strength.

Because no lateral pressure is employed in the unconfined compression test, it has several features: [6]

1. It is the simplest and quickest laboratory method commonly used to measure the shear strength of a cohesive soil.

2. It is used only on cohesive soils, since a cohesionless soil will not form an unsupported cylinder.

3. It is normally employed only for the second of the two approaches described above, i.e., as a measure of in situ strength.

---

[1] There is one uncommon exception to this general statement. A clay that in the past has been subjected to loads much greater than those currently applied may tend to expand upon shear. The prevention of drainage throws tension in the water, thus increasing the effective intergranular stress and, therefore, increasing the shear strength. Such a clay may have a greater undrained than drained shear strength when sheared under relatively small applied loads.

[2] The terms "slow" and "quick" come from the respective speeds at which drained and undrained shear tests are usually run.

[3] Although neither of these terms is well chosen, they will both be used in this book because of their general acceptance. The determination of sensitivity is discussed on page 118.

[4] Shear strength is sometimes presented as a function of water content or void ratio. See page 132 and Fig. XIII–11.

[5] The ratio of compressive to shear strength equal to two is based on a combination of theoretical and empirical considerations. See page 161, Appendix B.

[6] The unconfined test may give misleading results on heterogeneous soils because the lack of lateral support may be too severe a boundary condition. A check test in the triaxial machine is desirable for heterogeneous soils.

The unconfined compression test has an advantage over the direct shear because of the more uniform stresses and strains imposed. The pattern of strains shown in Figs. XI–1 and XI–2 is applicable to the comparison. Another advantage of the unconfined test is that the failure surface will tend to develop in the weakest portion of the clay. In contrast, the direct shear test forces the clay to shear along a predetermined surface, which may not be the weakest one.

The unconfined compression test, often called a U test, is a test at natural water content. (See introduction to Chapter XIII.) It is currently employed more than the other shear tests on cohesive soils. Portable unconfined compression tests are now used in this country and abroad for strength determinations at construction sites.

Apparatus and Supplies

Special

1. Unconfined compression device
2. Specimen trimmer with accessories
   (a) Miter box
   (b) Wire saw
   (c) Knives
3. Remolding cylinder and plunger

FIGURE XII–2. Unconfined compression machine. (Courtesy of Soil Testing Services, Inc., Chicago, Ill.)

General

1. Membrane for remolding
2. Balances (0.01 g sensitivity and 0.1 g sensitivity)
3. Drying oven
4. Desiccator
5. Timer

10. Evaporating dish
11. Wax paper (or cellophane)

As in other types of shear test apparatus, both stress control and strain control are used. Figure XII–2 shows a stress-controlled unconfined compression device; Fig. XII–4 shows a strain-controlled device. In the device shown in Fig. XII–2, the axial load is ap-

FIGURE XII–3. Unconfined compression machine. (Courtesy of Soil Testing Services, Inc., Chicago, Ill.)

6. Watch glasses or moisture content cans
7. Scale
8. Protractor
9. Spatula

plied by the horizontal bar, which is drawn downward by means of a jack. The large vertical wheel at the front controls the jack; the platform scale measures the load applied to the specimen.

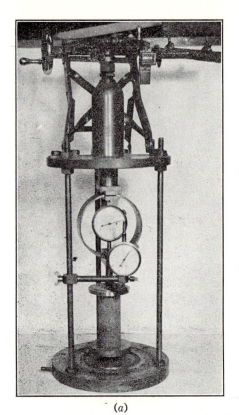
(a)

(b)

FIGURE XII–4. Unconfined compression test.

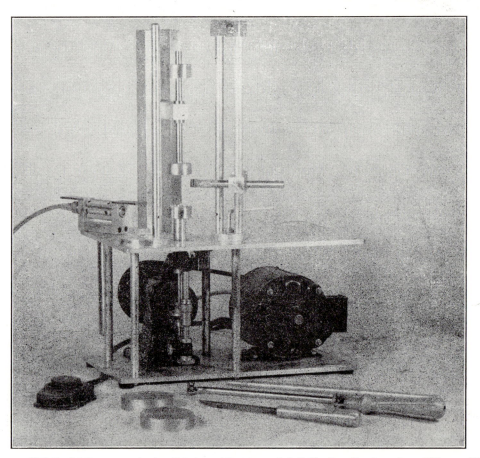

FIGURE XII–5. McRae motorized soil lathe. (Courtesy of Soil Testing Services, Inc., Chicago, Ill.)

In the machine shown in Fig. XII–3,[7] the load is applied to the specimen by a hydraulic system which is actuated by gas stored in the pressure drum at the right; the magnitude of the load is measured by the double proving ring.[8] This machine can be used in the field, since it is portable.

The apparatus in Fig. XII–4 is converted from a triaxial machine by removing the lucite chamber, using shorter vertical rods, and attaching an extensometer on one of the vertical rods.

Figure XII–5 shows a motorized soil lathe and trimming tools; Fig. XII–6 shows a miter box and a hand-operated soil trimmer. The trimmer in Fig. XII–5, which will trim either 1.4 in. or 2.8 in. diameter specimens, employs both a wire saw and a knife edge. The hand-operated device in Fig. XII–6 is able to trim to any of three diameters; it works in the same way as

3. Place the sample in the lathe [11] or trimmer (see Figs. XII–5 and XII–6b) and trim it to a circular cross section.

4. Again put the sample in the miter box and cut it to a length of approximately 3.5 in. by trimming both ends.

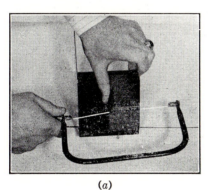

(a)

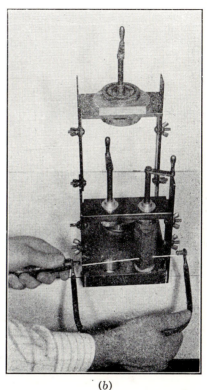

(b)

FIGURE XII–6. Trimming test specimen.

the trimmer used for the consolidation specimen. (See Fig. IX–7.)

Recommended Procedure [9]

Preparation of Undisturbed Specimen. Page 5 should be reviewed before the specimen preparation is started.

1. Cut from the clay supply a sample approximately 4½ in. long and 2 in. in diameter.[10]

2. Use the miter box and wire saw (or knife) to trim enough clay to make the ends parallel to each other by making them perpendicular to one of the sides. (See Fig. XII–6a.)

5. From the trimmings, obtain three representative specimens for water content determinations (one from near the mid height and one from near each end).

6. Measure the length and circumference of the specimen.

Compression Test. The detailed instructions which follow are based on a machine similar to that shown in Fig. XII–4. If another type is used, minor modifications in the procedure are required.

1. Place the specimen in the testing machine with its vertical axis as near the center of the loading plates as possible.

2. Obtain initial readings on the proving ring dial, timer, and vertical deflection dial, and then start the compression.[12]

---

[7] Approximate strain control can be obtained with the device shown in Fig. XII–3 by careful control of the pressure valve.

[8] See Appendix A, page 150.

[9] This test can be done better by two or more students. Students performing their first test should be able to make runs on an undisturbed specimen and a remolded specimen in about 2 hours, and to do the computations in about 2 hours. They will need supervision for the start of the test.

[10] The size of trimmed test specimen described in this test is 3.5 in. by 1.4 in. in diameter. See page 116.

[11] A motorized lathe can save trimming time. For soils which are difficult to trim (such as nonhomogeneous soils), however, a hand-operated trimmer may have to be resorted to for best results. The operator has better control of the cutting in the hand-operated one.

[12] A rate of strain of approximately ½% to 2.0% per minute is frequently employed for the strain-controlled test. An ap-

3. During early stages of the test, take readings [13] approximately every 0.01 in. of vertical deflection. As the stress-strain curve begins to flatten, take readings less often (i.e., every 0.02 in. and later every 0.05 in.). Figure XII–4a shows a specimen at the start of a test; Fig. XII–4b shows the sheared specimen.

4. Compress the specimen until cracks have definitely developed or the stress-strain curve is well past its peak.

5. Take the failed specimen into the humid room, and measure the angle between the cracks [14] and the horizontal. Sketch the failed specimen carefully. This sketch should be shown on the data sheet or, better, on the sheet presenting the stress-strain plot (see Fig. XII–9).

Preparation and Test of Remolded Specimen

1. Wrap the failed specimen [15] in a sheet of rubber and thoroughly remold it with the fingers. The remolding of the undisturbed specimen and the forming of the remolded specimen described in the next step should be done carefully to minimize drying of the soil, since to have the undisturbed and remolded strengths at the same water content is desirable.

2. Taking care to entrap as little air as possible, work the remolded soil into a 1.4 in. diameter mold with a spatula; extrude the specimen from the mold by means of a piston. For soft clays, more than one attempt may be required to mold an acceptable specimen.

3. Square the ends, measure the specimen, and test it as was done with the undisturbed specimen. However, readings may be taken less often (e.g., start with readings every 0.02 in. axial compression, changing later to every 0.05 in.). Compress until the stress-strain curve is well past its peak; if no peak is reached, stop the test at a strain of 20%.[16]

4. After the remolded test has been run, quickly place the specimen in a dish and weigh.

5. Sketch the failed specimen, and then put it in the oven. After it has dried and cooled, reweigh to determine the moisture content of the entire remolded specimen.

Discussion of Procedure

An unconfined specimen of homogeneous soil would be expected to fail near the mid height, since the ends of the specimen receive lateral restraint from the loading plates. However, a really homogeneous specimen of sedimentary clay is unusual (see Fig. I–5). Failure through the ends of the specimen (see Fig. XII–4b), which is not uncommon, indicates a zone of weakness. The ability to fail a specimen in a weak zone is one of the features of the unconfined test discussed on page 112.

Test specimens cut from tube samples are generally loaded parallel to the length of the sample because of size requirements. The common procedure, also, is to load specimens cut from pit samples in this vertical direction, which is usually approximately perpendicular to the stratification layers of a sedimentary soil (see Figs. I–5b and I–5c). There are indications that the strength is not very dependent on the direction of force application relative to stratification layers for some soils, but the opposite is true for others; this point should be checked on soils involved in important projects.

As with triaxial test specimens (see page 104), unconfined specimens should have a length-to-diameter ratio of 1.5 to 3.0; specimens having a ratio of 2½ are common. The specimen size (3.5 in. by 1.4 in. in diameter) employed in the recommended procedure of this book is used in many soil laboratories. There are indications,[17] however, that the strength of large masses of certain clays (e.g., brittle ones) may be a fraction of that indicated by tests on such small test specimens.

For a large program, only a limited number of tests on remolded specimens may be necessary because of constancy of the remolded strength of the clay. In this case a person may either partially dry his entire undisturbed specimen after the shear test in order to study stratification (see Fig. I–5b) or use it for a water-content determination. On the other hand, he may want to run Atterberg limits on the clay from the sheared specimen. Such decisions as these depend on the particular test program at hand.

---

plication of one-tenth to one-fifteenth of the estimated strength every ½ to 1 minute is common in the stress-controlled test. See page 127.

[13] See footnote 14, page 93.

[14] The angle between the horizontal and the failure plane should actually be measured at the time the failure plane is formed. Although a trained technician can probably measure this angle when it is formed, a student should not attempt to do so. Unless the soil specimen is homogeneous, the angle is not of much value.

[15] The student may find it desirable to add a little soil from the trimmings of the undisturbed specimen. By doing this, he will have enough clay to prepare a remolded specimen 3.5 in. long, and thus can use Fig. XII–7 in his computations.

[16] The selection of 20% is arbitrary.

[17] Rutledge found for one brittle lacustrine clay full of small slickensided surfaces that the strength of a large mass was 55% to 60% of that obtained from compression tests on 1.4 in. diameter specimens.

Occasionally the unconfined specimen is coated with grease (e.g., Vaseline), heavy oil, or a rubber membrane to retard surface drying. The use of grease or oil is open to question when the soil will be reused for Atterberg limits or a compression test on a remolded specimen unless the grease or oil is completely removed. At any rate, the time during which the trimmed specimen is exposed to the normal humidity before completion of the test should be minimized. Also, the ends of unconfined specimens are sometimes capped in plaster of Paris to insure better bearing. This capping is desirable when the soil is of such a nature (e.g., friable or stony) that smooth square ends cannot be prepared.

Calculations

The axial strain, ϵ, is found from

$$\epsilon = \frac{\Delta L}{L_0} \qquad (XI-2)$$

in which ΔL = change of specimen length as read from an extensometer,

L_0 = the initial specimen length.

The average cross-sectional area, A, can be found from

$$A = \frac{A_0}{1 - \epsilon} \qquad (XI-3)$$

in which A_0 = the initial area of the specimen.

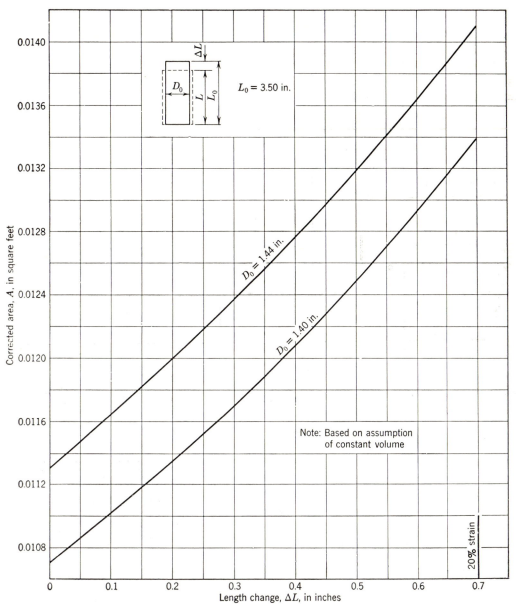

FIGURE XII-7. Area versus change of length.

Plots of A against ΔL for 1.40 in. diameter and 1.44 in. diameter by 3.5 specimens are given in Fig. XII–7.

The shear stress, τ, is taken equal to one-half the compressive stress, f_c, and can be found from the equation [18]

$$\tau = \frac{f_c}{2} = \frac{P}{2A} \qquad \text{(XII–1)}$$

in which P = compressive force = (proving ring reading − initial proving ring reading) × ring calibration factor.[19]

1. Divide the peak undisturbed strength by the remolded strength obtained at a magnitude of strain equal to that at which the peak undisturbed strength occurred.

2. Divide the peak undisturbed strength by the peak remolded strength.

If no peak strength occurs (see Fig. XII–9, for example), the strength at some arbitrarily defined strain, such as 15% or 20%, is used for the peak. In this book, the first method is used.

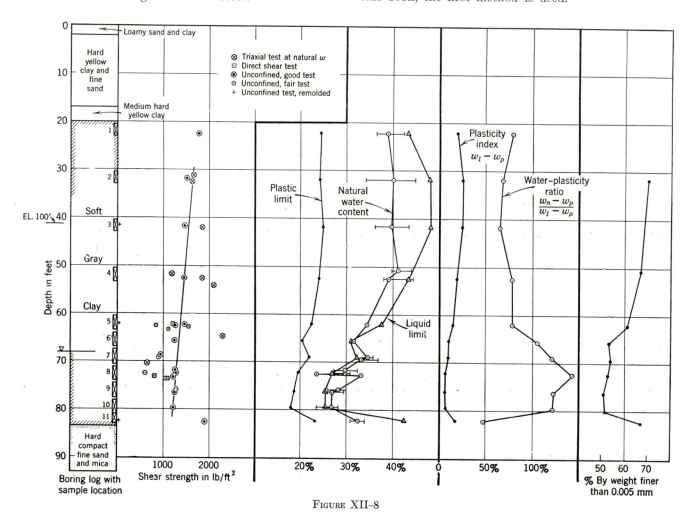

Figure XII–8

As stated on page 111, the loss [20] of strength of a cohesive soil due to remolding is called the sensitivity of the soil, which is defined as the undisturbed strength divided by the remolded strength. Among the methods of computing sensitivity are:

[18] Shear stresses based on the initial area, A_0, instead of A, are sometimes reported.

[19] See footnotes 15 and 16 on page 94.

[20] We should not infer that the strength of a soil after remolding is necessarily less than that of the undisturbed soil. A soil which is remolded and then consolidated may have

The slope of the initial portion of the stress-strain curve, known as the modulus of deformation, is sometimes computed. When a continuously flattening

greater strength. At a pressure greater than the maximum precompression pressure, a remolded specimen would exist at a lower void ratio than it would if undisturbed. In other words, remolding prior to consolidation to a large pressure results in a more dense soil. The strength increase from this increased denseness may offset the structural strength of the soil, with the result that the remolded soil has a greater strength than the undisturbed.

curve is obtained and it is concluded that the slope of the initial tangent is not significant, the modulus may be taken as the slope of a straight line from the origin to a point on the stress-strain curve. This point can be selected arbitrarily or to mark the limit of the stress range which is of interest.

Results

Method of Presentation. The results of an unconfined test can be presented in a summary table and/or by a stress-strain curve. A table might give for the undisturbed and remolded specimens the peak and ultimate values of $P/2A$, along with the strain at which each occurs. Also the water content of both specimens and the sensitivity should be included. A stress-strain curve usually consists of $P/2A$ or P/A versus strain for both specimens. Sketches of the failed specimens (see Fig. XII–9) should be shown on the stress-strain plots.

In an investigation of in situ strength (see page 111) the test results are plotted as a function of depth. Figure XII–8 is an example of such a plot. In Fig. XII–8, results from the identification or classification tests are also plotted against depth in order to check for relationships between the strength and these results. From a correlation with other test results, we may also be able to detect plausible explanations for unusual strength results. For such a study, a summary plot, as Fig. XII–8, is very useful and is often employed in foundation investigations for important structures.

In addition to unconfined test results, the results of both direct shear and triaxial tests are plotted in Fig. XII–8. The scattering of strength values displayed in Fig. XII–8 is not uncommon for investigations on heterogeneous clays. In an attempt to aid the interpretation of the strength tests, each unconfined test plotted in Fig. XII–8 was classified as good or fair. This classification was based on the stress-strain

curve, appearance of the sample from which the specimen was taken, type of failure, appearance of partially dried specimen, etc. As a consequence, those tests which do not appear typical are given less weight.

Typical Values. An indication of typical values of shear strength is given by the following classification (XII–1) of clay based on consistency:

Consistency of Clay	Shear Strength ($\frac{1}{2}$ compressive strength in lb/sq ft)
Very soft	<250
Soft	250–500
Medium	500–1000
Stiff	1000–2000
Very stiff	2000–4000
Hard	>4000

A value of sensitivity greater than 3 or 4 is considered high. Since high sensitivity means that a large portion of the strength of the soil is lost on remolding, we should be very cautious about using such a soil in a structure where remolding or progressive action is likely to occur.

Numerical Example

In the example on pages 120 and 121, complete test data are given only for the undisturbed specimen. A summary of results from both the undisturbed and the remolded test is:

	Specimen	
	Undisturbed	Remolded
$P/2A$ peak in lb/sq ft	1590	.....
$\Delta L/L_0$ peak in %	2.0	.....
$P/2A$ ultimate in lb/sq ft	1380	180
$\Delta L/L_0$ ultimate in %	4.3	20.0
w in %	38.8	38.5
Sensitivity	80	

REFERENCE

1. Terzaghi, Karl, and Ralph Peck, *Soil Mechanics in Engineering Practice,* John Wiley and Sons, New York, 1948.

SOIL MECHANICS LABORATORY

UNCONFINED COMPRESSION TEST

SOIL SAMPLE _Silty Clay: gray at natural water content, inorganic; glacial origin, sedimentary deposit; extremely sensitive; medium plastic; very soft when remolded._

SOIL SPECIMEN MEASUREMENTS

CIRCUMFERENCE _4.52 in._
INITIAL AREA, A_0 _0.0113 sq. ft._
INITIAL LENGTH, L_0 _3.50 in._

TEST NO. _U-17_
DATE _Feb. 8, 1950_
TESTED BY _WCS_

LOCATION _Union Falls, Maine_
BORING NO. _GA_   SAMPLE DEPTH _26.9 ft._
SAMPLE NO. _GA-3- 26.9_
SPECIFIC GRAVITY, G_s, _2.75_

PROVING RING NO. _12_
CALIBRATION FACTOR _0.5 lb. per 0.0001 in._

WATER CONTENT

SPECIMEN LOCATION	ENTIRE REMOLDED SPECIMEN		TOP	MID	BOTTOM
CONTAINER NO.	D 23		E-11	E-15	E-8
WT. CONTAINER + WET SOIL IN g	253.9		17.306	17.504	17.279
WT. CONTAINER + DRY SOIL IN g	209.4		14.666	14.704	14.561
WT. WATER, W_w, IN g	44.5		2.640	2.800	2.718
WT. CONTAINER IN g	93.6		7.835	7.503	7.553
WT. DRY SOIL, W_s, IN g	115.8		6.831	7.201	7.008
WATER CONTENT, w, IN %	38.5		38.6	38.9	38.8

ELAPSED TIME IN min.	VERTICAL DIAL IN in.	STRAIN, ε, IN %	AREA, A, IN sq. ft.	PROVING RING DIAL IN .0001 in.	AXIAL LOAD, P, IN lbs.	SHEAR STRESS, P/2A, IN lbs./sq. ft.
0	1.000	0	0.0113	0	0	0
	0.990	0.3	0.0114	9.0	4.50	198
	.980	0.6	0.0114	21.0	10.50	462
	.970	0.9	0.0114	38.0	19.00	580
	.960	1.1	0.0114	53.2	26.60	1163
	.950	1.4	0.0115	64.5	32.25	1405
	.940	1.7	0.0115	71.2	35.60	1550
	.930	2.0	0.0115	73.3	36.65	1590
	.920	2.3	0.0116	73.5	36.75	1588
	.910	2.6	0.0116	71.6	35.80	1545
	.900	2.9	0.0116	70.1	35.05	1505
	.890	3.1	0.0116	69.0	34.50	1477
	.880	3.4	0.0117	67.9	33.95	1450
	.870	3.7	0.0117	66.9	33.45	1425
	.860	4.0	0.0118	65.8	32.90	1400
6.0	0.850	4.3	0.0118	65.0	32.50	1380

REMARKS:

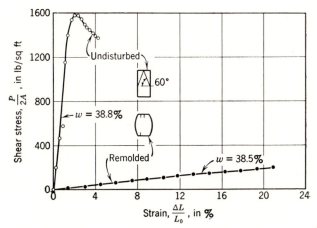

FIGURE XII–9. Unconfined compression test.

XIII

Triaxial Compression Test on Cohesive Soil

Introduction

In Chapter XII there was a brief discussion of the fundamentals of shear strength of cohesive soils, in which drained or slow shear and undrained or quick shear were described. Drained shear, in which the excess pore water pressure is negligible, and undrained shear, in which no escape of pore water is permitted, comprise the limiting conditions of drainage.

Not only is the nature of the actual shear process important, however, but the condition of the soil prior to shear is also significant. From the discussion of consolidation in Chapter IX, it follows that either of the two initial states is possible.

1. The soil may be completely consolidated to the pressure system to which it is subjected.

2. The soil may be only partially consolidated to the pressure system to which it is subjected.

Combining these two possible initial conditions of consolidated and unconsolidated with the two extremes of drainage mentioned in the first paragraph of this chapter gives three possible limiting-type shear tests. They are:

1. Consolidated—undrained, also called consolidated-quick or Q_c, for short.
2. Unconsolidated—undrained, also called unconsolidated-quick or Q, for short.
3. Drained, also called slow or S, for short.

A special type of number 2 is one in which the soil is sheared at its natural water content.

As has been pointed out already, the frictional resistance of a soil depends directly on the pressure acting between the particles, or intergranular pressure. A specimen, therefore, will exhibit greater strength when tested in drained shear than when tested in undrained shear.[1] By similar reasoning a higher strength will be obtained by permitting the specimen to consolidate prior to shear. Not only is the existing pressure to which a cohesive soil is consolidated important but also its pressure history is important. If a soil has been consolidated, in its past, to a pressure which is greater than the pressure to which it is now consolidated, it is a precompressed or preconsolidated soil (see page 83). A cohesive soil has greater strength if it has been precompressed. As can be seen from Fig. IX–12, a soil at a given intergranular pressure is denser after precompression than it was before and is, therefore, stronger. In addition, precompression often gives the soil "structural" strength. This strength comes from an attraction between particles which is probably due to some alteration of the water films on the particles. The general statements above are illustrated by Fig. XIII–9, which shows plots of data for a series of each type of test described.

The unconfined compression test described in the preceding chapter is an unconsolidated-undrained, or Q, test. Although some slight drainage may occur during the test, it is essentially undrained; and, since there are no means of applying a lateral pressure, a specimen cannot be consolidated in the unconfined apparatus prior to shear.[2] On the other hand, all three tests, Q, Q_c, and S, can be run in the direct shear and triaxial machines. In this chapter the triaxial

[1] There is an uncommon exception to this statement. See footnote 1 on page 111.

[2] Sometimes unconfined tests are run on specimens which have been consolidated by capillary forces obtained by permitting desiccation.

test is presented, and in the following chapter the direct shear test is presented. Also in Chapter XIV the relative merits of the triaxial versus the direct shear machines for tests on cohesive soils are discussed. This discussion points out that the triaxial is more controllable and dependable.

For a Q test the triaxial machine shown in Fig. XI-3 can be used as it is. However, for a Q_c or S test, to have facilities for top as well as bottom drainage is desirable. These can be obtained by running an outlet tube from the cap vent to a burette in a fashion similar to the base drainage outlet. For the

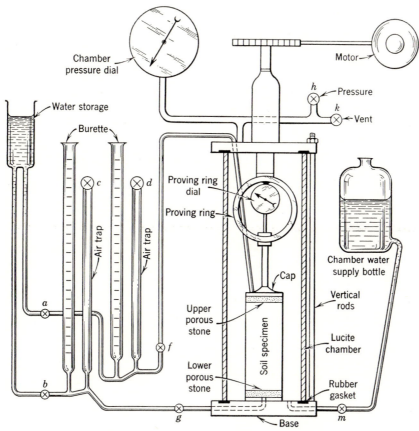

FIGURE XIII-1. Triaxial compression apparatus.

Apparatus and Supplies

Special

1. Triaxial machine
2. Specimen trimmer with accessories
    (*a*) Miter box
    (*b*) Wire saw
    (*c*) Knives
3. Rubber membrane
4. Membrane stretcher

General

1. Deaired water supply
2. Vacuum supply
3. Balances (0.1 g and 0.01 g sensitivity)
4. Drying oven
5. Desiccator
6. Rubber strips for binding
7. Evaporating dishes
8. Timer

testing of clays which are not completely saturated, the insertion of bubble traps in the drainage tubes is desirable. In Fig. XIII-1 is a sketch of the apparatus with these alterations made. For triaxial testing of clays, recessed caps and bases have been used (at Harvard University) to force the failure of the soil specimen to occur away from the ends.

The sample trimmers shown in Figs. XII-5 and XII-6b can be used to prepare triaxial specimens. In Fig. XIII-2 is shown a triaxial specimen being trimmed in the same device used for preparing the unconfined specimen (Fig. XII-6b). The cutting, however, is being done with a knife made from a straight-edge razor, the blade of which is slightly tilted toward the specimen.

A membrane stretcher is a cylindrical tube which is larger in diameter than the soil specimen and has a tube connected to the side. Figure XIII-3 shows a sketch of a stretcher with a membrane in place on it.

Recommended Procedure [3]

Preparation of Undisturbed Specimen. As pointed out in Chapter I, the specimen should be prepared in a humid room, with every reasonable precaution taken to minimize drying and disturbance of the soil. Review page 5 before starting the specimen preparation. Two specimen sizes [4] widely used for

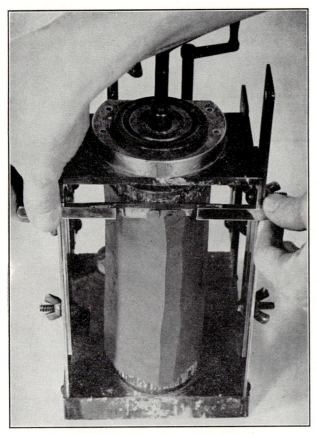

FIGURE XIII–2. Trimming a triaxial specimen.

triaxial tests are 6.5 in. in height by 2.8 in. in diameter and 3.5 in. in height by 1.4 in. in diameter. The trimming procedure used is identical with that given on page 115 for the unconfined specimen if a 3.5 by 1.4 in. specimen is to be prepared. The only alterations in this procedure when a 6.5 by 2.8 in. specimen is to be prepared are the obvious ones in the dimensions called for. Figure XIII–2 shows a 6.5 by 2.8 in. triaxial specimen being trimmed with a knife. After the specimen is trimmed, measure its length and circumference.

[3] This test can be done better by two or more students. They should be able to run their first Q test in about 3 hours and do the computations in 2 to 3 hours. They will need supervision for most of the test.

[4] A ratio of specimen length to diameter of 1.5 to 3 is recommended; a ratio of 2½ is common.

Compression Test (see Fig. XIII–1)

1. While in the humid room, enclose the specimen within the membrane by using a membrane stretcher. This is done by lapping the ends of the membrane over the stretcher and then applying suction through the tube (see Fig. XIII–3). The membrane and the stretcher are easily slid over the specimen, the suction released, and the membrane unrolled from the ends of the stretcher.[5]

2. Weigh the membrane-enclosed specimen to 0.1 g.

For Q_c and S tests:

3. Deair the base and the line connecting the base to the burettes by flushing with boiling water. Place a porous stone, which has been boiled in water to remove the air, in the base of the machine.

4. Allowing water to run out of the base slowly to prevent entrapping air, place the specimen on the bottom porous stone. Next moisten the membrane, overlap, and bind [6] it to the base with a rubber strip.

5. Place the cap with the upper porous stone on top of the specimen. During this step, water should be running slowly through the cap to prevent entrapping air between the stone and soil. As with the base, the cap and line connecting the cap to the burette should have been deaired.

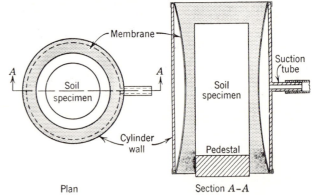

Plan   Section *A–A*

FIGURE XIII–3. Membrane stretcher.

For Q tests:

3a. Since no escape of the pore water is to be permitted, nonporous plates can be used instead of porous stones.

[5] A metal or a plastic cap can be temporarily placed on the ends of the specimen to keep the edges of the specimen from being disturbed when the membrane is unrolled from the stretcher. These caps are removed just before the specimen is placed in the triaxial machine.

[6] The binding should be done very carefully. The successive laps should be put close together but not on top of each other; at least five or six laps of well-stretched rubber should be made. Leakage of pore water may occur through careless binding.

4a. Place the specimen on the bottom plate, moisten the membrane, overlap, and bind it to the base with a rubber strip.

5a. Place the cap on top of the specimen.

For all tests:

6. Moisten the upper end of the membrane, and roll it over the sides of the cap.

7. Screw in one vertical rod, attach a clamp from the cap to the rod, thus lining up and supporting the cap, and carefully bind the membrane to the cap with a rubber strip.

8. Remove the clamp, then screw the remaining vertical rods in the base.

9. Next wet the bottom rubber gasket, center the lucite chamber [7] on this gasket, wet the upper rubber gasket, and place it on top of the lucite cylinder.

10. Carefully put the upper assembly of the machine in place, then check to see that the tip of the plunger contacts the center of the cap.

11. Tighten [8] all the top nuts of the vertical rods until they just begin to bind, and then give each one-fourth revolution turn. Keep giving each nut one-fourth turn until two complete turns have been given.

12. By means of a length gage as a check, further tighten any nuts as far as necessary to make the upper plate parallel to the base.

13. At this point all valves (a to m inclusive) should be closed except the vent k.

14. Admit water [9] to the chamber by opening valve m. Bring the water level up until the cap is just covered, then close valve m.

15. Close valve k, open the pressure drum, set the desired chamber pressure (see page 126) on the regulator. Build up the chamber pressure slowly by cracking valve h.

16. Bring the plunger down until it is just in contact with the cap. Contact is indicated by a deflection of the proving ring dial.

For Q_c and S tests:

17. Permit the specimen to consolidate [10] under the applied chamber pressure by opening both top and bottom drainage valves (g, f). The quantity of water expelled during consolidation can be measured in the burettes and the quantity of air in the traps.

18. After consolidation is complete (completion is indicated by the stoppage of the flow of water into

the burettes), again bring the plunger into contact with the cap. The distance the plunger is moved is the change of length of the specimen due to consolidation.

For Q_c tests:

19. Close the drainage valves (g, f).

For all tests:

20. At this stage, carefully check to see that all is ready to begin testing. A student should check with his instructor.

21. Record the initial proving ring dial, start the motor, and, when the proving ring dial begins to move, engage the motor revolution counter and start the timer.

22. Take readings of counter, proving ring dial, burettes (for S test), and time.

23. A set of readings [11] should be taken every 0.01 in. of compression for the first 0.05 in., and then approximately every 0.02 for 0.1 in. for undrained tests; readings are needed less frequently in drained tests.

24. Throughout the test make sure that the chamber pressure is held constant. The pressure can be increased by increasing the pressure on the drum regulator and released through valve k. Readings of time need to be taken only every third or fourth reading. After the peak point is passed, readings may be taken less frequently.

25. Continue the test until the applied compressive force remains constant for a few readings or until the specimen has been strained approximately 15%.[12]

26. Stop the compression and release the axial load. Close valve h and then slowly release the chamber pressure through valve k.

27. Drain the chamber by lowering the supply bottle and opening valve m.

28. Disassemble the apparatus (removing the chamber nuts one-half or one revolution at a time).

29. Sketch the failed specimen; on the sketch dimension the maximum and minimum diameters, the length of the specimen and the angle of inclination of the shear plane, if there is one.

30. Remove the membrane enclosed specimen from the machine and weigh to 0.1 g. The difference between this weight and that obtained in step 2 is the amount of water lost through drainage, and can be checked against the drainage determined from burette readings.

31. If a test on a remolded specimen is desired, remold (see page 116), measure the specimen, and re-

[7] See footnote 17, page 103.

[8] A torque wrench is often employed to control the tightening.

[9] See page 129 for other chamber fluids used.

[10] Consolidation to principal stress ratios other than unity can be obtained more conveniently with stress-controlled loading units than with strain-controlled ones.

[11] See footnote 14, page 93.

[12] The selection of 15% as a maximum strain is arbitrary.

peat the compression test. For very soft soils it is easier to remold the undisturbed specimen within the membrane and then shape it with a mold. Less frequent readings are necessary during the shear of the remolded specimen.

test, the nature of the soil, and the information desired from the test results. If the first approach, described on page 111, of relating shear strength to pressure (or water content or void ratio) is to be used, a series of tests [13] at different chamber pressures is re-

Figure XIII–4. Setup for measuring pore water pressure during a triaxial test.

32. Otherwise, remove the membrane from around the specimen, and place the specimen in the oven. After drying and cooling it, reweigh for a water content determination. (If partially dried specimens are desired for a study of stratification [see Fig. I–5c], only a portion of the specimen should be used for the water content determination.)

Discussion of Procedure

Magnitude of Chamber Pressure. The magnitude of the chamber pressure which should be used in a triaxial test on cohesive soil depends on the type of

quired. An example of the relation between boundary pressure and shear strength for the three types of tests (Q, Q_c, and S) can be obtained from Fig. XIII–9.

To get a measure of the in situ strength of the soil either of the following can be used:

[13] Sometimes a strength envelope can be obtained from a single test (reference XIII–7) by varying the chamber pressure when the specimen has just reached failure. Such a test, known as a multiple-stage test, is particularly useful for strength studies on partially saturated soils which possess little structure, such as compaction specimens. The data employed in Fig. XIII–10 were obtained by multiple-stage tests.

1. A Q test with the chamber pressure equal to the overburden pressure that existed on the soil sample in nature.

2. A Q_c test with the chamber pressure equal to the intergranular pressure that existed on the soil sample in nature.

The second of the foregoing tests is thought to give the better indication of in situ shear strength.

Special Test. A special type of triaxial test [14] deserves consideration. This is an undrained triaxial test in which the pore water pressure is measured [15]

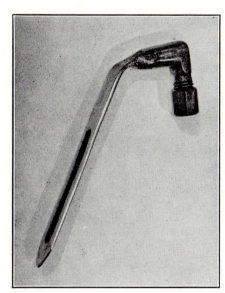

FIGURE XIII–5. Porous pilot for measuring pore water pressure.

during shear by means of a porous pilot inserted near the center of the specimen. The pilot is connected to a capillary tube outside the triaxial chamber by a plastic tube. The air pressure required to maintain the water level at the calibration mark in the capillary tube is equal to the water pressure in the soil pores existing at the location of the pilot. Knowing the magnitude of the pore water pressure during shear, we can compute the intergranular stresses and thus determine the true strength envelope. In other words, we are able to secure the information from an undrained test which we would normally be able to determine only from a drained test.

Figure XIII–4 shows a porous pilot, capillary tube, and gage for measuring the pore water pressure; Fig. XIII–7 shows the pilot installed. Figure XIII–5 shows a close-up of a porous pilot which consists of

[14] For a more detailed description of this test, see reference XIII–10.

[15] See reference XIII–4 for a more detailed description of this method of measuring pore water pressures.

three thicknesses of screen, one No. 100 and two No. 150, within a slotted, flattened brass tube.

Taylor (XIII–6) feels that the undrained test with pore pressure measurements is the most dependable method of determining the shear strength of a cohesive soil. It is his standard method of shear strength determination. The Bureau of Reclamation (XIII–8) uses a triaxial test in which pore water pressures are measured at the specimen ends. Although use of triaxial tests with pore pressure measurements is increasing, the higher degree of technical skill needed to perform it will probably retard its acceptance.

Rate of Shear. The strength of a cohesive soil generally increases as the rate of shear is increased. For example, Casagrande found from undrained tests that the strength of a very soft organic clay when sheared in 1.7 minutes was 40% greater than when sheared in 7 hours (XIII–1). The effect of rate of shear on strength is further illustrated in Fig. XIII–6, which is a plot of compressive strength determined by undrained tests against rate of axial strain for a remolded Boston blue clay. Several clays, when failed in dynamic tests, in which only 0.02 second elapsed between the start of shear and the attainment of maximum compressive stress, showed strengths 1.4 to 2.6 times those obtained with a 10-minute loading time (XIII–2).

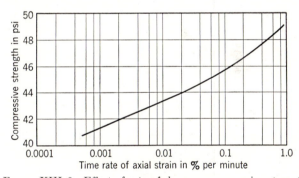

FIGURE XIII–6. Effect of rate of shear on compressive strength of remolded Boston blue clay ($w = 29\%$). (From reference XIII–10.)

The results above indicate that the strength of a clay may be significantly affected by extreme rates of shear. In a practical problem, when employing values of clay strength determined in the laboratory, we must consider the relationship of strength versus rate of shear. Although careful control of rate of shear is not usually necessary for laboratory tests, an approximate standard is desirable in order to have a basis for comparison of the strengths of different soils.

For undrained tests on stress-controlled equipment, a load application of approximately one-fifteenth of the compressive strength every ½ minute has been recommended (XIII–11). For undrained tests on strain-controlled equipment, rates of axial strain from ½% to 1% per minute are often used. For drained tests using continuous loading, the rate of loading must be sufficiently low to prevent any pore water pressures from developing. As an example, a rate of

be large enough to allow complete consolidation to take place before the next load increment is applied. As an example, one day is sufficient time for a specimen of undisturbed Boston blue clay, 6.5 in. by 2.8 in. in diameter, to consolidate under an increment of load.

FIGURE XIII–7. Extensometer inside triaxial chamber.

axial strain less than 0.06% per hour was necessary to prevent building up pore water pressure in a specimen of undisturbed Boston blue clay, 6.5 in. by 2.8 in. in diameter (the permeability of this clay is approximately 2×10^{-7} cm/second). For drained tests in a stress-controlled apparatus, load increments of a certain fraction of the compressive strength are applied at regular time intervals. These intervals should

Measurement of Specimen Deformation. As stated on page 104, an extensometer for measuring specimen deformation can be mounted within the chamber of the triaxial machine. With such an installation we can measure the deformation directly, rather than compute it from the number of revolutions which the drive shaft makes. To take direct measurements within the chamber involves practical difficulties; however, if these are overcome, dependable readings can be obtained. Figure XIII–7 shows an extensometer mounted for direct measurement of deformation of a triaxial specimen.

Chamber Fluids. In the procedures presented in this chapter and in Chapter XI, water was used as a chamber fluid to apply lateral pressure to the specimen. Although water is satisfactory for tests of short duration—e.g., a couple of days or less—it is not recommended for longer tests because it tends to permeate the membrane. The same holds for air, nitrogen, or other gases which are sometimes used. Although not as convenient to handle as water or gas, a fluid [16] such as glycerin, castor oil, or olive oil should be used for the longer tests.

Membranes. Natural rubber is the best material for the thin membranes used to cover the specimens. Although the membranes should be impermeable to the chamber fluid and strong enough not to break during a test, they should not furnish restraint to the compression of the soil. Warlam (XIII–9) found that resistance to deformation of a thin rubber membrane added 10% to 20% to the strength of a soft Chicago clay. He found further that the membrane effect increased with chamber pressure and was higher for a shear failure than for a plastic flow failure.[17]

Calculations

For the strain-controlled apparatus used in the procedure described above, the change of specimen length, ΔL, is found from equation

$$\Delta L = a - b \qquad \text{(XI–1)}$$

where a = movement of the top of the proving ring = k (in inches per revolution) $\times$ number of revolutions. k is a property of the gear box and is found by calibration.

b = compression of proving ring = proving ring dial reading in inches − initial proving ring dial reading in inches.

The axial strain, ϵ, at any time is found from

$$\epsilon = \frac{\Delta L}{L_1} \qquad \text{(XIII–1)}$$

where L_1 = the sample length at the start of shear. (For the Q test, $L_1 = L_0$. For the Q_c or S test, $L_1 = L_0 - \Delta L_c$, where ΔL_c is the change of length from consolidation. See step 18.)

[16] Carbon tetrachloride has also been used; however, it seriously weakens a rubber membrane.

[17] See page 106 for a description of shear and plastic flow failure.

The deviator stress, p, is equal to the difference between the major and minor principal stresses, where the principal stresses are either combined or intergranular stresses, or

$$p = \sigma_1 - \sigma_3 = \bar{\sigma}_1 - \bar{\sigma}_3 \qquad \text{(XIII–2)}$$

The deviator stress can be computed from

$$p = \frac{P}{A} \qquad \text{(XI–4)}$$

where P [18] = applied force = (proving ring reading − initial proving ring reading) $\times$ ring calibration factor,[19]

A = the cross-sectional area of the specimen.

The average cross-sectional area, A, is computed from:

For Q tests:

$$A = \frac{A_0}{1 - \epsilon} \qquad \text{(XI–3)}$$

where A_0 = the initial area.

A plot of A versus ΔL for a specimen 2.8 in. in diameter by 6.5 in. is given in Fig. XIII–8.

For Q_c tests: [20]

$$A = \frac{A_c}{1 - \epsilon} \qquad \text{(XIII–3)}$$

where A_c = the average area of the specimen following the initial consolidation and can be found from equation [21]

$$A_c = \frac{V_0 - \Delta V_c}{L_0 - \Delta L_c} \qquad \text{(XIII–4)}$$

where V_0 = initial specimen volume,

ΔV_c = volume change during consolidation = volume of water drained during consolidation for saturated soil,

ΔL_c = length change during consolidation (see step 18),

L_0 = initial specimen length (i.e., as trimmed).

For S tests:

$$A = \frac{V_0 - \Delta V}{L_0 - \Delta L} \qquad \text{(XIII–5)}$$

[18] See footnote 15, page 94.

[19] See footnote 16, page 94.

[20] A more precise expression (reference XIII–11) is $A = \dfrac{A_c}{1 - C\epsilon}$,

where $C = \dfrac{A_{em}}{A_{e\epsilon}}$, in which A_{em} is the measured area at the end of test and $A_{e\epsilon}$ is the computed area at the end of test.

[21] An equation (reference XIII–11), sometimes used instead of Eq. XIII–4, is $A_c = A_0 \dfrac{L_0 - 2\Delta L_c}{L_0}$.

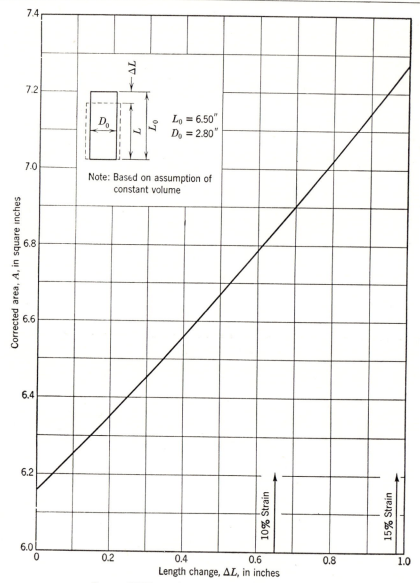

FIGURE XIII–8. Area versus change of length.

The ratio of major and minor principal intergranular stresses, $\bar{\sigma}_1/\bar{\sigma}_3$, cannot be computed from the data of the normal Q or Q_c test. In the S test, however, $\bar{\sigma}_1/\bar{\sigma}_3$ can be computed from Eq. XIII–6 because the pore water pressures are equal to zero.

$$\frac{\bar{\sigma}_1}{\bar{\sigma}_3} = \frac{\dfrac{P}{A} + \bar{\sigma}_3}{\bar{\sigma}_3} = \frac{\dfrac{P}{A} + \sigma_{ch}}{\sigma_{ch}} \qquad \text{(XIII–6)}$$

in which σ_{ch} = chamber pressure.

If $\bar{\sigma}_1/\bar{\sigma}_3$ is known, the true friction angle,[22] ϕ, can be computed from

$$\phi = \sin^{-1} \frac{(\bar{\sigma}_1/\bar{\sigma}_3) - 1}{(\bar{\sigma}_1/\bar{\sigma}_3) + 1} \qquad \text{(XI–6)}$$

In an undrained test with pore pressure measurements the ratio $\bar{\sigma}_1/\bar{\sigma}_3$ can be computed as shown below.

$$\bar{\sigma}_3 = \sigma_3 - u = \sigma_{ch} - u \qquad \text{(XIII–7)}$$

where u is the pore water pressure, and

$$\bar{\sigma}_1 = (\bar{\sigma}_1 - \bar{\sigma}_3) + \bar{\sigma}_3 = \frac{P}{A} + \bar{\sigma}_3 \qquad \text{(XIII–8)}$$

The friction angle,[22] ϕ, can be computed from Eq. XI–6 because $\bar{\sigma}_1/\bar{\sigma}_3$ is known.

For either a drained test or an undrained test with measured pore water pressures, the angle θ between the

[22] The friction angle, ϕ, in a cohesive soil can be computed only from a specimen which is not precompressed. This point is illustrated in Fig. XIII–9.

horizontal (actually the major principal plane) and the failure plane at the time of its formation can be computed from

$$\theta = 45° + \frac{\phi_m}{2} \qquad (XI-7)$$

The methods of computing the sensitivity and modulus of deformation described on page 118 for the unconfined test are also applicable to the triaxial test. If no peak strength occurs, the strength at some arbitrarily defined strain, such as 15%, is used for the computation of sensitivity.

Results

Method of Presentation. The results of a triaxial compression test on cohesive soil can be presented by a stress-strain curve and/or in a summary table. The data of Q and Q_c tests are usually presented as $\sigma_1 - \sigma_3$ plotted against axial strain. In an S test $\bar{\sigma}_1 - \bar{\sigma}_3$, $\bar{\sigma}_1$, or $\bar{\sigma}_1/\bar{\sigma}_3$, along with volume change, can be plotted against axial strain. A summary of results might give the peak and ultimate strength along with the strain at which each occurs. The water content of the test specimen should also be given.

The method of presenting the results of a test program depends on the choice of the two approaches described on page 111. A plot of strength versus depth (see Fig. XII–8) is helpful to the study of the in situ strength of the soil at a particular site. Either Q or Q_c tests (see page 127) can be used to determine the strength. The type of plot used to present the results of a series of tests as a function of pressure depends on

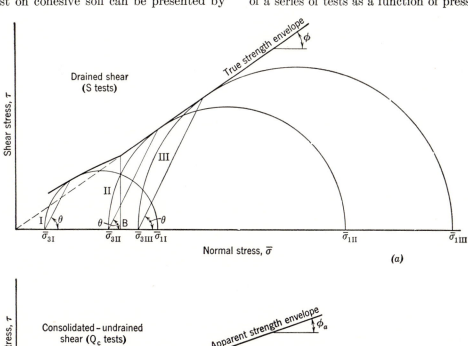

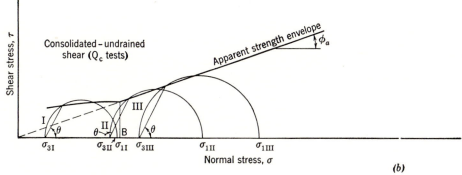

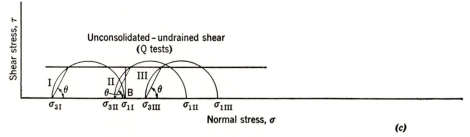

FIGURE XIII–9. Triaxial compression tests on a saturated clay.

the type of test run. Figure XIII–9 shows plots of
three series of tests on a saturated cohesive soil which
has been precompressed to pressure B.

Figure XIII–9a shows the Mohr circles for three S
tests and the strength envelope for the test series.
Since the water pressure is maintained at zero in S
tests, all pressures are intergranular. The circle for
each test is drawn through $\bar{\sigma}_1$ and $\bar{\sigma}_3$ at failure, with
the diameter of the circle being the compressive strength
$\bar{\sigma}_1 - \bar{\sigma}_3$. The strength envelope is drawn through the
failure points. The envelope at pressures greater than
B is tangent to the Mohr circles and intersects the origin if extended, as shown by the dashed line in Fig.
XIII–9a. The failure point on the circle is determined
by a line making an angle $\theta = 45° + (\phi/2)$ with the
major principal plane. The slope of the straight portion of the true envelope is the true friction angle, ϕ.
At any normal stress the difference in shear stress between the envelope and the dashed line is cohesion.

Figure XIII–9b shows the Mohr circles for three
tests in which the specimens were initially consolidated to the same pressure as the correspondingly
numbered test in Fig. XIII–9a and then sheared with
no drainage (Q_c tests). Since the tests are undrained,
all normal stresses are water plus intergranular
stresses. The circle for each test is drawn through
σ_1 and σ_3 at failure, with the diameter of the circle
being the compressive strength $\sigma_1 - \sigma_3$. The strength
envelope is drawn through the failure points which
are located on the circles by lines making an angle
$\theta = 45° + (\phi/2)$ with the major principal plane (see
Fig. XIII–9b). A value of 60° for θ is usually close
enough to be used for locating the failure point on the
circle. The slope of the straight portion of the apparent envelope is the apparent friction angle, ϕ_a.

Figure XIII–9c shows the Mohr circles for the three
tests in which no consolidation either before or during
shear is permitted (Q tests). The normal stress in
Fig. XIII–9c is water plus intergranular pressure.
The circles and envelopes in Fig. XIII–9c are drawn
by the same procedure as those in Fig. XIII–9b. Since
an increase in lateral pressure is carried by the pore
water and, therefore, generates no additional strength,
the envelope is horizontal.

If the soil is partially saturated, i.e., contains air, an
increase of lateral pressure brings about a decrease in
volume even though no drainage is permitted. The
volume decrease is caused by compression of the entrapped air. Because of decreased soil volume, the
strength of a partially saturated soil is increased by
increases in lateral pressure. Thus the envelope for
a series of Q tests on a partially saturated soil is
not horizontal like that for the saturated soil (Fig.

XIII–9c), but slopes upward for higher pressures.
This sloped envelope is illustrated in Fig. XIII–10
(reference XIII–7) which shows undrained shear envelopes for samples of a given deposit at various degrees of saturation. Figure XIII–10 shows a greater
effect of lateral pressure on those samples having the
lowest percentage of saturation.

There are good indications that the strength of a
cohesive soil is a constant function of void ratio, or
water content for a saturated soil, at failure regard-
less of the type of test or test condition (XIII–5,

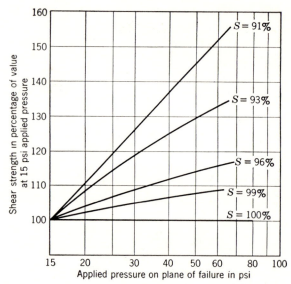

FIGURE XIII–10. Strength envelopes for a partially saturated
clay.

XIII–11). Because of this the results of a shear program are often presented in plots of strength on a log
plot versus void ratio. Such a plot appears to be an
approximately straight line and parallel to the virgin
compression curve of the soil as determined from a
consolidation test (see Chapter IX). The relationship
described above is illustrated by Fig. XIII–11, where
a plot of compressive strength versus void ratio at
failure is placed on a standard consolidation test plot.

If the results of a shear test with known pore pressures are available, there is another way (XIII–6) to
indicate shear strength. A plot of shear strength divided by the major principal intergranular stress is
made (strength divided by minor principal intergranular stress is sometimes used); from such a plot
the shear strength can be obtained for any known
major principal intergranular stress. This type of
plot is illustrated in Fig. XIII–12b, which is a plot
of data from the Numerical Example.

Typical Values. Typical values of the shear
strength of clay were given on pages 118 and 119. A

typical value of ϕ is approximately 30°, whereas ϕ_a is often in the neighborhood of half of ϕ. We can also obtain an indication of shear strengths from the various plots given in this chapter.

In Fig. XIII–12 the test data are plotted. The usual curve for an undrained test is darker in Fig. XIII–12a; the usual plot of an S test is shown darker in Fig. XIII–12b. In addition to the values previously de-

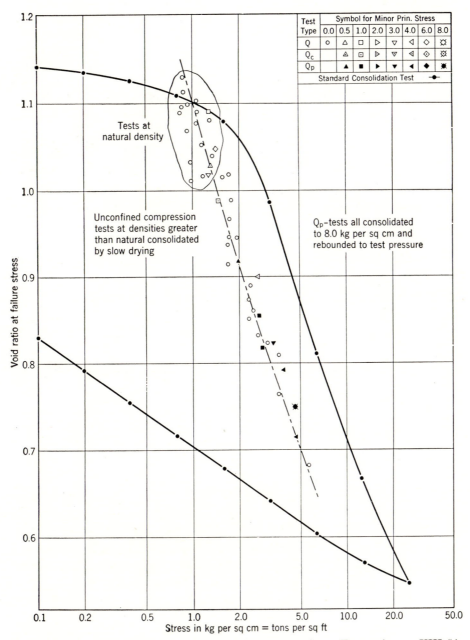

FIGURE XIII–11. Stress versus void ratio for a Minnesota clay. (From reference XIII–5.)

Numerical Example

For the numerical example (pages 135, 136, and 137) of a triaxial shear test on a clay, a Q_c test with pore water pressure measurements has been selected. Not only does this test illustrate the computations and plots for a normal Q or Q_c test, but it indicates how much information can be gained from a knowledge of pore water pressures.

scribed, the shear stress on the plane on which the obliquity angle (the angle between the resultant force and a tangent to the plane) is at a maximum is plotted. For strains less than that at which this shear stress, τ_{cr},[23] is maximum, the shear strength, s, is taken as the

[23] $\tau_{cr} = \dfrac{(\bar{\sigma}_1 - \bar{\sigma}_3)\sqrt{\bar{\sigma}_1\bar{\sigma}_3}}{\bar{\sigma}_1 + \bar{\sigma}_3}.$

maximum value of τ_{cr}. At any strain larger than this strain, s is taken as equal to the τ_{cr} occurring at the given strain.

The results of the example are summarized below:

P/A peak in psi	36.6
ϵ at peak P/A in %	2.6
τ_{cr} peak in psi	16.6
ϵ at peak τ_{cr} in %	1.6
P/A ultimate in psi	23.2
ϵ at ultimate P/A in psi	13.8
τ_{cr} ultimate in psi	9.0
ϵ at ultimate τ_{cr} in %	13.8
ϕ_m in degrees	35.5
ϕ_u in degrees	35.5
w in %	32.1

REFERENCES

1. Casagrande, A., and W. L. Shannon, "Stress-Deformation and Strength Characteristics of Soils under Dynamic Loads," *Proceedings of the Second International Conference on Soil Mechanics,* Vol. 5, Paper II d 10, p. 29, 1948.
2. Casagrande A. and W. L. Shannon "Research on Stress-Deformation and Strength Characteristics of Soils and Soft Rocks under Transient Loading," Publication from Harvard Graduate School of Engineering, No. 447, *Soil Mechanics Series,* No. 31, 1947–1948.
3. Hvorslev, M. J., "Torsion Shear Tests and Their Place in the Determination of the Shearing Resistance of Soils," *Proceedings of the American Society for Testing Materials,* Vol. 39, 1939.
4. Lambe, T. W., "The Measurement of Pore Water Pressures in Cohesionless Soils," *Proceedings of the Second International Conference on Soil Mechanics,* Paper II a 9, Vol. VII, June, 1948.
5. Rutledge, P. C., "Strength of Natural Clays," presented at the October, 1948, meeting of the American Society of Civil Engineers.
6. Taylor, D. W., "Shearing Strength Determinations by Undrained Cylindrical Compression Tests with Pore Measurements," *Proceedings of the Second International Conference on Soil Mechanics,* Vol. V, p. 45, 1948.
7. Taylor, D. W., "A Triaxial Shear Investigation on a Partially Saturated Soil," presented at the June, 1950, meeting of the American Society for Testing Materials.
8. Wagner, A. A., "Shear Characteristics of Remolded Earth Materials," presented at the June, 1950, meeting of the American Society for Testing Materials.
9. Warlam, Arpad A., "Stress-Strain and Strength Properties of Soils," Doctor of Science Thesis, Harvard University, May, 1946.
10. Waterways Experiment Station, "Shear Reports Made by M.I.T. Soil Mechanics Laboratory to U. S. Corps of Engineers," Vicksburg, Miss., Nos. 1–10, June, 1940, to May, 1944.
11. Waterways Experiment Station, "Triaxial Shear Research and Pressure Distribution Studies on Soils," *Progress Report,* Soil Mechanics Fact Finding Survey, Vicksburg, Miss., April, 1947.

SOIL MECHANICS LABORATORY

TRIAXIAL COMPRESSION TEST ON COHESIVE SOIL

SOIL SAMPLE _Silty Clay; grayish blue at natural water content; inorganic; glacial origin; sedimentary deposit; sensitive medium plastic; soft when remolded._

LOCATION _Cambridge, Mass._

BORING NO. _——_ SAMPLE DEPTH _7_

SAMPLE NO. _B-2_

SPECIFIC GRAVITY, G_s, _2.78_

SOIL SPECIMEN MEASUREMENTS

CIRCUMFERENCE _8.78 in._

INITIAL AREA, A_o _6.16 sq. in._

INITIAL LENGTH, L _6.50 in._

PROVING RING NO. _5_

CALIBRATION FACTOR _Calibration curve used._

TEST NO. _T-3_

DATE _March 17, 1950_

TESTED BY _VDB_

CHAMBER PRESSURE

DESIRED LATERAL PRESSURE _____ lbs./sq. ft. _54.9_ lbs./sq. in.

GAGE CORRECTION IN lbs./sq. in. _0.6_

GAGE PRESSURE IN lbs./sq. in. _55.5_

WATER CONTENT

SPECIMEN LOCATION	TOP	MIDDLE	BOTTOM	ENTIRE SPECIMEN, START OF TEST	ENTIRE SPECIMEN, END OF TEST
CONTAINER NO.	G-3	G-20	G-22	Membrane & covers	Membrane & covers
WT. CONTAINER + WET SOIL IN g	23.860	28.850	38.390	1671.85	1633.40
WT. CONTAINER + DRY SOIL IN g	21.005	24.840	31.900	1339.25	1339.25
WT. WATER, W_w, IN g	2.855	4.010	6.490	332.60	294.15
WT. CONTAINER IN g	13.275	13.085	13.591	422.16	422.16
WT. DRY SOIL, W_s, IN g	7.730	11.755	18.309	917.09	917.09
WATER CONTENT, w, IN %	37.0	34.1	35.4	36.3	32.1

DATE	TIME	ELAPSED TIME IN min.	COUNTER	CHAMBER PRESSURE IN lbs./sq. in.	PORE PRESSURE IN lbs./sq. in.	BOTTOM DRAINAGE IN cc WATER	BOTTOM DRAINAGE IN cc AIR	TOP DRAINAGE IN cc WATER	TOP DRAINAGE IN cc AIR	
March 17	4:00 P.M.	0	46000	55.5	52.5	23.8	3.85	19.2	1.15	
	4:20		46050	55.8	45.0	18.4	3.85	14.9	1.15	
	5:06		46059	55.5	33.0	15.2	3.85	11.7	1.15	
	5:06		"	"	"	24.0	3.85	22.6	1.15	*
	6:55		46072	55.5	18.0	20.1	3.85	18.5	1.15	
	10:00 P.M.		46081	55.5	7.6	17.3	3.85	15.5	1.15	
March 18	9:00 A.M.		46092	55.5	—	14.3	3.85	12.8	1.15	
	9:00		"	"		21.8	3.85	19.5	1.15	*
	4:00 P.M.		46095	55.5	0.9 **	21.0	3.85	18.9	1.15	
March 19	9:55 A.M.		46098	55.5	0.8 **	20.3	3.85	18.4	1.15	
	2:45 P.M.	2805	46098.5	55.5	0.5 **	20.1	3.85	18.3	1.15	

DRAINAGE IN cc 20.0 + 18.5 = 38.5 cc
= 2.35 cu. in.

$A_c = \dfrac{V_o - \Delta V_c}{L_o - \Delta L_c} = 5.89 \text{ in.}^2$

$L_c = L_o - \Delta L_c = 6.50 - 0.10 = 6.40 \text{ in.}$

REMARKS: * Water taken from burette.
** Measured with differential mercury manometer.

SOIL MECHANICS LABORATORY

TRIAXIAL COMPRESSION TEST ON COHESIVE SOIL

ELAPSED TIME IN min.-sec.	CHAMBER PRESSURE OBSERVED (lbs./sq. in.)	CHAMBER PRESSURE CORRECTED	COUNTER	PROVING RING DIAL IN .0001 in.	LENGTH CHANGE, ΔL, IN in.	STRAIN, ε, IN %	AREA, A, IN sq. in.	AXIAL LOAD, P, IN lbs.	AXIAL PRESSURE, $p=\frac{P}{A}$, IN lbs./sq. in.	PORE PRESSURE OBSERVED (lbs./sq. in.)	PORE PRESSURE CORRECTED	$\bar{\sigma}_3$ IN lbs./sq. in.	$\bar{\sigma}_1$ IN lbs./sq. in.	$\dfrac{\bar{\sigma}_1}{\bar{\sigma}_3}$	τ_{cr} IN lbs./sq. in.
0-0	55.5	54.9	46000	0	0	0	5.89	0	0		0.2 *	54.7	54.7	1	0
	"		20	96	.003	0.05	5.89	103.7	17.6		3.1 *	51.8	69.4		
	"		30	127	.006	0.09	5.89	135.7	23.0		6.8 **	48.1	71.1	1.48	11.3
	"		50	164	.022	0.35	5.91	175.5	29.7		9.3 **	45.6	75.3		
	"		60	174	.031	0.48	5.92	185.6	31.3		12.6 **	42.3	73.6	1.74	15.1
3-0	"		100	197	.071	1.11	5.96	209.5	35.2	22.0	21.4	33.5	68.7	2.05	16.5
	"		120	201	.092	1.44	5.97	213.8	35.8	27.0	26.0	28.9	64.7	2.24	16.5
	"		150	204	.123	1.92	6.01	216.9	36.2	30.3	29.4	25.5	61.7		
	"		165	205	.138	2.16	6.02	218.0	36.2	31.5	30.6	24.3	60.5	2.49	16.4
	"		180	206	.153	2.39	6.04	219.1	36.3	32.5	31.5	23.4	59.7		
	"		200	206	.174	2.72	6.06	219.1	36.2	33.5	32.5	22.4	58.6	2.62	16.2
	"		220	205.8	.194	3.03	6.08	218.8	36.0	34.0	33.0	21.9	57.9		
6-20	"		240	205.2	.215	3.36	6.10	218.2	35.8	36.0	34.9	20.0	55.8	2.79	15.8
	"		260	204.8	.236	3.69	6.12	217.7	35.5	36.5	35.4	19.5	55.0		
	"		280	203.7	.257	4.02	6.14	216.1	35.2	37.7	36.4	18.5	53.7	2.91	15.4
	"		300	202.5	.278	4.35	6.16	215.3	35.0	38.0	36.9	18.0	53.0		
8-20	"		320	201.8	.299	4.67	6.17	214.6	34.8	38.4	37.2	17.7	52.5	2.97	15.1
	"		340	200.8	.320	5.00	6.20	213.6	34.4	39.5	38.3	16.6	51.0		
9-25	"		360	199.6	.341	5.33	6.22	212.0	34.1	39.0	37.7	17.2	51.3	2.98	14.8
	"		380	198.9	.361	5.65	6.24	211.0	33.8	39.5	38.3	16.6	50.4		
10-25	"		400	197.8	.382	5.96	6.27	207.8	33.2	40.0	38.7	16.2	49.4	3.05	14.3
	"		420	197.1	.403	6.29	6.29	207.1	32.9	40.8	39.5	15.4	48.3		
	"		440	195.8	.424	6.61	6.31	205.8	32.6	41.2	39.9	15.0	47.6	3.16	13.9
11-29	"		460	195.1	.444	6.94	6.33	205.1	32.4	41.7	40.4	14.5	46.9		
	"		480	194.5	.465	7.27	6.36	204.1	32.1	41.9	40.6	14.3	46.4	3.25	13.6
	"		500	193.8	.486	7.60	6.38	203.5	31.9	42.1	40.8	14.1	46.0		
	"		520	192.8	.507	7.92	6.41	201.8	31.4	42.1	40.8	14.1	45.5	3.23	13.3
13-35	"		540	192.0	.528	8.25	6.43	201.0	31.2	42.1	40.8	14.1	45.3		
	"		560	191.5	.549	8.58	6.45	200.5	31.0	41.7	40.4	14.5	45.5	3.14	13.3
	"		580	190.8	.570	8.91	6.47	199.8	30.8	43.0	41.7	13.2	44.0		
	"		600	189.8	.591	9.23	6.48	198.8	30.7	43.2	41.9	13.0	43.7	3.36	12.9
17-45	"		640	188.3	.633	9.89	6.53	197.3	30.2	43.5	42.2	12.7	42.9		
	"		680	185.8	.675	10.55	6.58	194.3	29.5	43.5	42.2	12.7	42.2	3.32	12.4
	"		720	181.5	.716	11.20	6.64	190.0	28.6	43.5	42.2	12.7	41.3		
19-51	"		760	173.5	.758	11.85	6.68	182.0	27.2	44.5	43.1	11.8	39.0	3.31	11.5
	"		800	164.7	.799	12.50	6.73	172.7	25.6	46.1	44.5	10.4	36.0		
	"		840	153.2	.842	13.17	6.78	160.2	23.6	47.9	46.5	8.4	32.0	3.81	9.6
	"		880	152.1	.885	13.83	6.84	158.6	23.2	47.5	46.1	8.8	32.0		
26-08	55.5	54.9	47000	149.0	1.010	15.80	6.99	155.5	22.3	46.5	45.1	9.8	32.1	3.27	9.4

REMARKS:

* Read with manometer, therefore no gage correction.

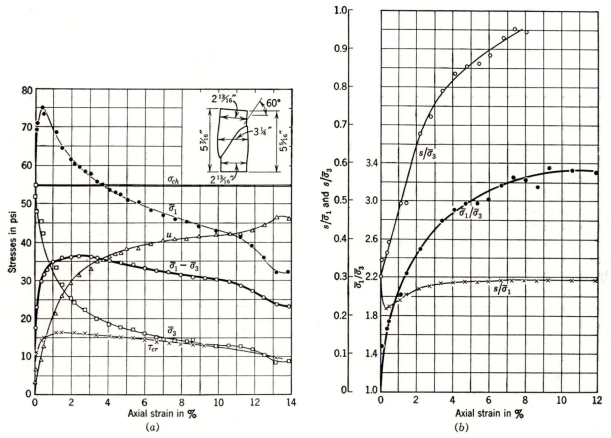

FIGURE XIII–12.   Triaxial compression test.

XIV

Direct Shear Test on Cohesive Soil

Introduction

In both Chapters XII and XIII a laboratory method of determining the shear strength of a cohesive soil was presented. In Chapter XII it was pointed out that, although the unconfined compression test has much merit for determining the strength of a soil at its existing water content, it cannot be used for tests in which specified pressure systems are desired. In other words, the unconfined compression test can be used only for Q tests and not for S or Q_c tests. The triaxial test which was presented in the last chapter and the direct shear test which is presented in this chapter can be used for all three types of tests. It appears desirable, therefore, to compare the triaxial and the direct shear tests with regard to the determination of the strength of cohesive soils.

The relative advantages and disadvantages of the triaxial and direct shear tests for sands presented in Chapter XI are applicable for the testing of cohesive soils (see footnote page 106). To the disadvantages listed for the direct shear test must be added the high degree of remolding of the soil which it fosters. This remolding tends to reduce substantially the test value of the ultimate strength of a clay.[1] For example, although the peak strength of undisturbed Boston blue clay as determined by Q_c direct shear tests agreed with that determined by Q_c triaxial tests, the direct shear ultimate strength was less than half that of the triaxial (XIV–1).

In the direct shear test, as has already been mentioned, the soil is failed along a plane predetermined

by the apparatus. For example, when the shear box and direct shear machine for which the detailed procedures are given in this book are used, the soil specimen is ruptured along a horizontal plane at mid height. On the other hand, in the unconfined and triaxial tests, the soil can fail within a zone of lower strength, should one exist. In the unconfined test (see Fig. XII–9) of the Numerical Example in Chapter XII, failure occurred at the top of the specimen, which indicates that it was weaker there than at the bottom.

The direct shear has an advantage over the triaxial test because less time is required for specimen drainage. Since the distance the escaping pore water has to flow is smaller for the direct shear specimen, less time is required for consolidation prior to shear in Q_c and S tests; likewise, the rate of shear in drained tests can be larger in the direct shear test than in the triaxial test. This faster drainage, in addition to the simpler and more rapid operation of the direct shear machine, makes the comparison of the required testing times for the triaxial and direct shear machines more favorable for the direct shear machine in testing cohesive soils than in testing cohesionless soils.

The complete prevention of drainage in the direct shear test is very difficult. For Q tests, nonporous gratings can be used in the soil container, but porous stones must be used for Q_c tests in order that the preshear consolidation may occur in a reasonable time. In the following pages an attachment to the direct shear machine to prevent drainage during shear is described. The great majority of so-called undrained tests in direct shear machines, however, are not completely undrained; unfortunately, there is no easy way to tell the degree to which these tests approach undrained ones. The ultimate strength is affected more than the maximum strength by this unknown quantity of drainage in Q and Q_c tests, because

[1] The fact that remolding may substantially reduce the ultimate strength can be used to obtain an indication of the susceptibility of the soil to progressive failure. For example, if a soil showed a much higher ultimate strength from a triaxial test than from a direct shear test, an indication of important progressive effects is given.

of the longer testing time required to reach the ultimate.

The foregoing discussion can be epitomized by the following statement: Although the direct shear machine is simpler and quicker than the triaxial machine for the shear testing of cohesive soils, it is neither as controllable nor as dependable as the triaxial. The simplicity and speed of operation of the direct shear machine give it a definite place in the shear testing of cohesive soils.

Apparatus and Supplies

Special

  1. Direct shear machine
  2. Tools for preparing specimen
    (*a*) Miter box
    (*b*) Wire saw and knife
    (*c*) Saw guides
    (*d*) Smooth surface (glass or plastic, etc.)
              or
  2*a*. Specimen trimmer with accessories
    (*a*) Wire saw or knife
    (*b*) Smooth surface (glass or plastic, etc.)

General

  1. Rubber sheet for remolding
  2. Balances (0.01 g sensitivity and 0.1 g sensitivity)
  3. Drying oven
  4. Desiccator
  5. Calipers
  6. Spatula
  7. Watch glasses or moisture content cans
  8. Timer
  9. Evaporating dish
  10. Wax paper (or cellophane)

Some of the differences in direct shear machines which are in common use were discussed in Chapter X in connection with sand testing. That discussion is generally applicable to direct shear testing on clays.

The machine (see Figs. X–8 and X–9) referred to in the procedure in Chapter X will also be adopted in the following recommended procedure. There are several modifications, however, which should be considered. The first modification is the replacement of the gratings (see Fig. X–9) with porous stones if either Q_c or S tests are to be run. The porous stones furnish exits for the escaping pore water. In S tests, prevention of evaporation is essential, and for this a large pan of water[2] which encompasses the entire shear box is desirable. Also recommended is the ad-

[2] Pads of cotton saturated with water can be placed around the shear box instead of the pan of water.

dition of a gear reduction box for a motorized machine to obtain the slow rate of shear needed.

In Fig. XIV–1 is shown a special device which has been used successfully to prevent drainage during a quick direct shear test. By operating the rotating crank during the test, we can keep the extensometer, which records the vertical deflection, in its initial position, thus maintaining a constant specimen height,

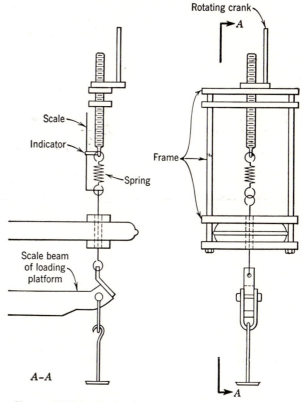

FIGURE XIV–1. Device for controlling specimen volume.

and presumably volume. The load applied to the beam is determined from the calibrated spring.

The wire saw, guides, and miter box used for trimming a direct shear specimen are shown in Fig. XIV–2. Also shown in Fig. XIV–3 is a trimmer which simplifies the cutting operation.

Recommended Procedure [3]

Preparation of Undisturbed Specimen. [4] Review page 5 before starting the specimen preparation.

[3] This test can be done better by two or more students. Students performing their first test should be able to make runs on an undisturbed and a remolded specimen in 2 hours, and do the computations in about 2 hours. They will need supervision at the start of the test.

[4] The direction of shear relative to the stratification in the soil which should be used in the laboratory tests depends on the particular soil and the use which is to be made of the re-

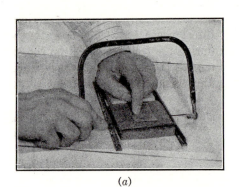 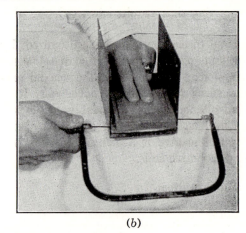

(a)   (b)

FIGURE XIV-2. Trimming a direct shear specimen.

FIGURE XIV-3. Direct shear specimen trimmer.

The procedure for preparing the test specimen will be given for both trimmers mentioned above. The size of the trimmed specimen used in the direct shear machine shown in Fig. X–8 is 3 in. by 3 in. by ½ in. thick.

Miter Box and Guides.

1. From the soil to be tested, trim a sample approximately 3½ in. by 3½ in. by ¾ in. thick, with one of the large faces cut carefully to a plane.

2. With the level face down, cut the sample ½ in. thick, using the saw guides as shown in Fig. XIV–2a.

3. Place the upper frame of the shear box lightly on one of the large faces of the sample and carefully trace on the sample the inside of the box with a knife.

4. Next cut the sample along the knife marks, using the miter box as shown in Fig. XIV–2b.

5. From the soil trimmings obtain at least three representative specimens (approximately 10 g each) for water content determinations.

sults. To shear the soil across the stratification instead of parallel is usually reasonable. On some soils the direction of shear is not important; on others it makes a great deal of difference.

Trimmer (Fig. XIV–3).

1. From the soil to be tested, trim a sample approximately $3\frac{1}{2}$ in. by $3\frac{1}{2}$ in. by $\frac{3}{4}$ in. thick, with one of the large faces cut carefully to a plane.

2. Place the sample in the trimmer with the level face on the grooved base of the trimmer.

3. If the soil is friable, put the square plate on the soil and trim to size, one edge at a time, with a knife (see the top of Fig. XIV–3).

4. If the soil is soft, trim it by slowly lowering the frame with the mounted cross wires (foreground of Fig. XIV–3) through the soil into the grooves of the base.

5. Trim the sample to the desired [5] thickness by drawing the wire saw horizontally along the top surface of the frame with the cross wires.

6. From the soil trimmings obtained, keep at least three representative specimens (approximately 10 g each) for water content determinations.

Shear Test on Undisturbed Specimen

1. Counterbalance the device used to apply the normal load or obtain the tare weight. The tare weight is the scale reading with the bottom half of the box on the platform.

2. Take the shear box into the humid room and place the prepared specimen in the shear box. Next bolt the upper frame to the lower and put the upper grating (or porous stone if Q_c or S test) and loading block in position.

3. Place the loaded shear box on the scale platform and bolt it into position. For an S test, attach the large pan which contains the shear box and fill the pan with water (see footnote 2, page 139).

4. Apply the desired normal load. For a Q_c or S test, allow time (see page 142) for consolidation. For an unconsolidated-undrained (Q) test, a normal load of approximately $\frac{1}{3}$ to $\frac{1}{2}$ T per sq ft is suggested (see page 142 for discussion of this suggestion). For a Q test, start shear as soon after applying the normal load as possible in order to minimize drainage.

5. Separate the parts of the soil container approximately $\frac{1}{40}$ in., tighten the bolts which hold the upper grating (or porous stone for Q_c or S test), set the extensometer dials, back off the spacer screws, and then remove the lock screws (see step 11, page 93). Check

to see that nothing but soil connects the two parts of the shear box. Take initial dial readings.

6. *For Q_c and Q test:* [6] Start the test, take readings of time, proving ring dial (shear force) and shear displacement dial at intervals of 15 seconds, for the first 2 minutes, then at every 0.020 in. of horizontal displacement (see footnote 14, page 93). Continue the test until shear force is constant for a few readings or to a shear displacement of 0.5 in.

6a. *For S test:* Start the test; take readings of time, proving ring dial, shear and normal displacement. The rate of shear displacement should be in the neighborhood of 2 to 3×10^{-4} in. per minute (the slower rate for the more plastic clays). Readings need not be taken as frequently as for undrained shear. Continue the test until the shear force is constant for a few readings or to a displacement of 0.5 in.

7. If no test on a remolded specimen is to be run, remove the shear specimen from the machine and weigh. Finally dry in the oven and reweigh for a water content determination.

Preparation of Remolded Specimen

1. Quickly remove the sheared undisturbed specimen from the machine, wrap it [7] in a rubber sheet, and remold it thoroughly with the fingers. This should be done in a humid room, with every precaution taken to prevent loss of moisture from the soil.

Shear Test on Remolded Specimen

1. Bolt the upper and lower frames of the shear box together. While in the humid room, work the remolded soil into the box, taking care to avoid entrapping air bubbles in the soil.

2. The actual testing of the remolded specimen is identical with that for the undisturbed, except that we need not take test readings as frequently.

3. At the conclusion of the test, quickly remove the sheared specimen, and weigh. Finally dry in the oven and reweigh for a water content determination.

Discussion of Procedure

Magnitude of Normal Load. As was the case in the selection of the chamber pressure for the triaxial test (see page 126), the magnitude of normal load which should be used in a direct shear test on cohesive soil depends on the type of test, nature of the soil, and the information desired from the test results. Tests

---

[5] The thickness can be altered by means of the two large adjusting screws on the frame (see Fig. XIV–3). For a Q_c or S test on a highly compressible soil, a larger trimmed thickness may be necessary in order to have a thickness of about $\frac{1}{2}$ in. at the start of shear.

[6] For complete prevention of drainage, a device such as that shown in Fig. XIV–1 to maintain the initial reading of the extensometer which records vertical deflection is suggested.

[7] To add a little clay from the trimming of the undisturbed specimen to make up for the soil unrecovered from the rubber sheet after remolding is desirable.

at different normal pressures will be required in order to make plots like those in Fig. XIV–5. As Fig. XIV–5c shows, the magnitude of the normal load is not important in a true Q test on a saturated soil. The $\frac{1}{3}$ or $\frac{1}{2}$ T per sq ft, suggested in step 4, is large enough to keep the specimen within the box during shear. Large loads are not recommended for undrained tests because of the difficulty of completely preventing drainage. The larger the applied load, the greater the tendency for drainage.

If we desire to get a measure of the in situ strength of a cohesive soil, either of the following tests can be made:

1. A Q test with the normal pressure equal to the in situ overburden pressure that existed on the soil sample.

2. A Q_c test with the normal pressure equal to the in situ intergranular pressure that existed on the soil sample.

The second of the two tests is thought to give the better measure of in situ shear strength.

Time Allowed for Consolidation. For Q_c or S tests, time should be allowed for complete dissipation of pore water pressures before the start of the shear test. By observing the vertical deflection dial, we can determine when consolidation has been completed because, at this point, compression will substantially cease (except on organic soils which exhibit a measurable amount of compression after the pore water pressures are negligible). Since 15 to 25 minutes is sufficient for consolidation of a $\frac{1}{2}$ in. specimen of most clays, 40 minutes is suggested as a conservative value. Taylor reported (XIV–1) that the shear strength of undisturbed Boston blue clay was dependent on the time the normal load was left on the specimen after consolidation had been completed. He found that if one day instead of 40 minutes was allowed for consolidation, he obtained approximately 10% greater strength. In view of such evidence, the amount of time allowed for consolidation should be standardized for a series of tests.

Rate of Shear (see page 127). Rates of displacement from 0.04 in. to 0.10 in. per minute give substantially the same strength for undrained direct shear tests. For drained shear tests, the rate must be sufficiently low to permit the escape of pore water. A rate in the neighborhood of 2 to 3×10^{-4} in. per minute has been found satisfactory for three inch square specimens of many clays. Approximately 2 days are required to shear a specimen to 0.5 in. at the rate of 2×10^{-4} in. per minute.

Calculations

The shear stress, τ, can be found from

$$\tau = \frac{F}{A} \qquad (XIV–1)$$

in which F[8] = shear force = (proving ring reading − initial ring reading) × ring calibration factor,

A = cross-sectional area of the specimen (taken as constant and equal to the initial area for routine testing).

The determination of the effective shear area, A, depends on the details of the soil container. If the box (see Fig. XIV–4) has a sharp lip, i.e., t is negli-

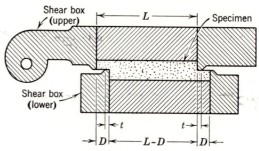

FIGURE XIV–4. Shear box.

gible, the area A is equal to $w(L − D)$, where w is the box width and L and D are as shown in Fig. XIV–4. On the other hand, if the box has a broad lip, i.e., $t = 0.2$ to 0.3 in., the effective area depends on the adhesion between the sheared soil and the lip. If we assume that this adhesion is half the shear strength of the soil,[9] then we obtain a constant area. The assumption of a constant area is probably a good one for small displacements like those required for the development of the peak strength on most undisturbed clays. An accurate determination of the effective area for circular boxes or rings becomes a complicated problem in geometry.

For a particular box, we can make a plot of effective area, A, against displacement, D. We must remember, however, that the most important advantage of the direct shear test is simplicity. Therefore, complicated expressions for A, and the resulting difficulty of maintaining a constant normal pressure in the drained test, materially weaken the arguments for using the direct shear machine.

The methods of computing the sensitivity and modulus of deformation described on page 118 for the unconfined tests are also applicable to the direct shear

[8] See footnotes 15 and 16 on page 94.
[9] There is some experimental justification for this assumption (XIV–1).

test. If no peak strength occurs, the strength at some arbitrarily defined shear displacement, such as 0.5 in., is used for the computation of sensitivity.

Results

Method of Presentation. The results of a direct shear test on a cohesive soil can be presented in a

The method of presenting the results of a series of shear tests depends on which of the two approaches described on page 111 is used. For studying the in situ strength of the soil at a particular site, a plot of strength versus depth is helpful (see Fig. XII–8). Either Q or Q_c tests (see page 142) can be used to determine the strength. If shear strength as a function

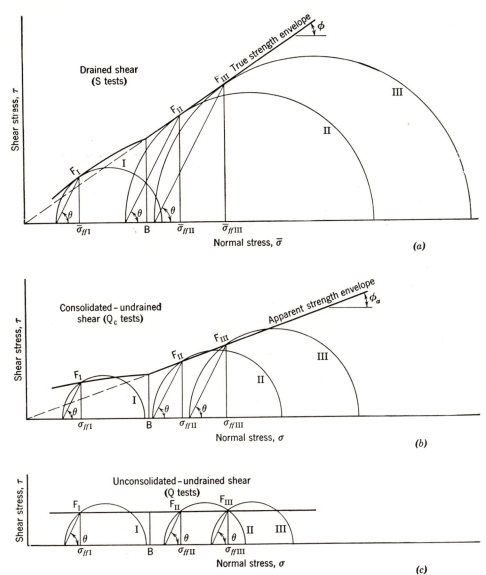

FIGURE XIV–5. Direct shear tests on a saturated clay.

summary table and/or by a stress-strain curve. A table might give for the undisturbed and remolded specimens the peak and ultimate shear strength along with the shear displacements at which each occurs. The water content of both specimens and the sensitivity of the sample are usually included. A stress-strain curve normally consists of shear stress versus shear displacement for both the undisturbed and the remolded tests.

of pressure is being studied, plots such as Fig. XIV–5 can be made. Figure XIV–5 shows plots of three series of tests on a saturated cohesive soil which has been precompressed to pressure B.

Figure XIV–5a shows the Mohr circles for three S tests and the envelope for the test series. Since the water pressure is maintained at zero in S tests, all pressures are intergranular. The envelope is drawn through the shear stress at failure, or shear strength (labeled F in Fig. XIV–5a), which is plotted at the normal stress on the failure plane at failure, $\bar{\sigma}_{ff}$. The slope of the straight portion of the true envelope is the true friction angle, ϕ. The circle for a test in which $\bar{\sigma}_{ff}$ is greater than the preconsolidation pressure (B) is drawn through the failure shear stress and tangent to the envelope. The circle for a test in which $\bar{\sigma}_{ff}$ is less than the preconsolidation pressure is drawn through the failure shear stress and the minor principal stress. The minor principal stress is determined by a line through the failure point and at an angle $\theta = 45° + (\phi/2)$ from the horizontal.

Figure XIV–5b shows the Mohr circles for three tests in which the specimens were initially consolidated to the same pressure as the correspondingly numbered test in Fig. XIV–5a and then sheared with no drainage (Q$_c$ tests). Since the tests are undrained, all normal stresses are water plus intergranular stresses. The strength envelope is drawn through the shear strength (labeled F in Fig. XIV–5b), which is plotted against the normal stress on the failure plane at failure, σ_{ff}. The pressure σ_{ff} is also the pressure to which the specimen was consolidated prior to shear.[10] The slope of the straight portion of the apparent envelope is the apparent friction angle, ϕ_a. The minor principal stress is next located by a line from the failure point F, which makes an angle of $\theta = 45° + (\phi/2)$ with the major principal plane. A value of 60° for θ is usually close enough for this location. The Mohr

circle for a test is now drawn through the minor principal stress and the failure point.

Figure XIV–5c shows the Mohr circles for three unconsolidated-undrained tests (Q tests) and the envelope for the series. Since no drainage is permitted, all normal stresses are combined stresses. The strength envelope and circles in Fig. XIV–5c are obtained in a similar manner to those in Fig. XIV–5b. Because the water carries any increase in normal pressure, and thereby prevents any more friction from being mobilized, the envelope in Fig. XIV–5c is horizontal.

As illustrated by Fig. XIII–10, an increase of pressure on a partially saturated soil causes an increase of strength in undrained shear.

Typical Values. Typical values of the shear strength of clay were given on page 119. A typical value of ϕ is approximately 30°, whereas ϕ_a is often in the neighborhood of half of ϕ.

Numerical Example

The example on pages 145 and 146 is an unconsolidated-undrained (Q) direct shear test on an inorganic clay from Maine. Only the test data for the test on the undisturbed specimen are given. The data, however, for tests on both undisturbed and remolded specimens are plotted in Fig. XIV–6 and included in the summary of results below.

	Specimen	
	Undisturbed	Remolded
s_m in lb/sq ft	808	......
Shear displacement at s_m in inches	0.051	......
s_u in lb/sq ft	513	79
Shear displacement at s_u in inches	0.480	0.500
w in %	34.3	33.3
Sensitivity	28	

[10] Actually the apparent strength envelope is a plot of shear strength against the pressure to which the specimen was consolidated prior to shear, whereas a Mohr circle is a representation of stresses within the specimen. The applied pressure is kept constant in the normal Q$_c$ test, and the initial consolidation pressure, therefore, is equal to the applied normal pressure at failure, σ_{ff}. For these normal tests, the envelope passes through the Mohr circles as shown in Fig. XIV–5$_c$. For those tests where the consolidation pressure is not equal to σ_{ff} (as would be true if the device shown in Fig. XIV–1 were used), the abscissas of the apparent envelope and stress circles are different. Then the apparent envelope does not intersect the Mohr circle.

REFERENCE

1. Waterways Experiment Station, "Shear Reports Made by M.I.T. Soil Mechanics Laboratory to U. S. Corps of Engineers," Vicksburg, Miss., Sept. 30, 1949.

SOIL MECHANICS LABORATORY

DIRECT SHEAR ON COHESIVE SOIL

SOIL SAMPLE _Sandy Silty Clay: gray at_
natural water content; inorganic; glacial origin;
sedimentary deposit; extremely sensitive;
medium plastic; very soft when remolded.
LOCATION _Union Falls, Maine_
BORING NO. _GG_ SAMPLE DEPTH _73.3 ft._
SAMPLE NO. _GG-8-73.3_
SPECIFIC GRAVITY, G_s, _2.75_
TYPE OF TEST _Q_

SOIL SPECIMEN MEASUREMENTS
LENGTH _____3 in._____
AREA, A_o, IN sq. ft. _.0625_
THICKNESS _____0.50 in._____

PROVING RING NO. _____1_____
CALIBRATION FACTOR _0.5 lb per .0001 in._

TEST NO. _D-39_
DATE _Sept. 19, 1950_
TESTED BY _WCS_

SCALE LOAD
APPLIED LOAD _41.7_ lbs. _667_ lbs./sq. ft.
TARE IN lbs. _43.0_
SCALE LOAD IN lbs. _84.7_

WATER CONTENT

SPECIMEN LOCATION	TOP	BOTTOM	EDGE	ENTIRE REMOLDED SPECIMEN	
CONTAINER NO.	E-11	E-15	E-8	D-18	
WT. CONTAINER + WET SOIL IN g	17.067	17.161	16.369	227.4	
WT. CONTAINER + DRY SOIL IN g	14.754	14.730	14.035	193.3	
WT. WATER, W_w, IN g	2.313	2.431	2.334	34.1	
WT. CONTAINER IN g	7.835	7.503	7.553	90.0	
WT. DRY SOIL, W_s, IN g	6.919	7.227	6.482	103.3	
WATER CONTENT, w, IN %	33.5	33.7	35.9	33.3	

ELAPSED TIME IN min.	SHEAR DIAL IN in.	SHEAR DISPLACEMENT IN in.	PROVING RING DIAL IN .0001 in.	SHEAR FORCE IN lbs.	SHEAR STRESS, τ, IN lbs./sq. ft.
0	0.0	0.0	0	0	0
1/4	0.005	.005	42.0	21.0	336
1/2	0.016	.016	75.0	37.5	600
3/4	0.033	.033	94.0	47.0	752
1	.051	.051	101.0	50.5	808
1 1/4	.070	.070	101.0	50.5	808
1 1/2	.092	.092	97.5	48.7	780
1 3/4	.115	.115	92.5	46.7	740
2	.138	.138	87.5	43.7	700
	.160	.160	83.5	41.7	668
	.180	.180	81.5	40.7	652
	.200	.200	80.2	40.1	642
	.220	.220	77.5	38.7	620
	.240	.240	76.8	38.4	615
	.260	.260	75.2	37.6	602
	.280	.280	73.2	36.6	585
	.300	.300	72.2	36.1	577
	.320	.320	71.2	35.6	570
	.340	.340	70.0	35.0	560
	.360	.360	68.8	34.4	550
	.380	.380	68.0	34.4	544
	.400	.400	67.0	33.5	536
	.420	.420	66.2	33.1	530
	.440	.440	65.5	32.7	524
	.460	.460	64.8	32.4	518
	.480	.480	64.2	32.1	513
6 3/4	.500	.500	64.2	32.1	513

REMARKS:
Rupture area assumed constant.

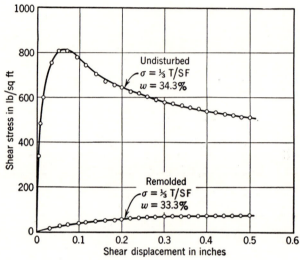

Figure XIV-6. Direct shear test.

APPENDIX A

TABLE A–1. CONVERSION FACTORS

Length

1 in. = 2.540 cm
1 ft = 30.480 cm
1 μ = 0.0001 cm = 10^{-4} cm
1 μ = 0.001 mm = 10^{-3} mm
1 mμ = 10^{-3} μ
1 Å = 0.1 mμ
1 Å = 1/10,000 μ = 0.0001 μ = 10^{-4} μ
1 Å = 0.0000001 mm = 10^{-7} mm
1 Å = 0.00000001 cm = 10^{-8} cm

Area

1 sq ft = 144 sq in.
1 sq cm = 100 sq mm
1 sq in. = 6.452 sq cm
1 sq ft = 929.03 sq cm

Volume

1 cu ft = 1728 cu in.
1 cu cm = 1000 cu mm
1 cu in. = 16.39 cu cm
1 cu ft = 28317 cu cm

Velocity

1 in. per sec = 2.540 cm per sec
1 ft per sec = 30.48 cm per sec
1 ft per min = 0.508 cm per sec
1 ft per min = 5080 μ per sec

Mass

1 lb = 453.6 g
1 kg = 2.2 lb

Pressure

1 atm = 14.7 psi
1 atm = 76.0 cm Hg (at 0° C)
1 atm = 33.9 ft water (at 4° C)
1 atm = 2116 lb per sq ft
1 psi = 6.895 × 10^4 dynes per sq cm
1 psi = 5.17 cm Hg (at 0° C) = 5.19 cm Hg (at 20° C)
 = 2.03 in. Hg (at 0° C) = 2.04 in. Hg (at 20° C)
1 psi = 70.29 cm water (at 4° C) = 70.43 cm water (at 20° C) = 27.67 in. water (at 4° C) = 27.73 in. water (at 20° C)
1 psi = 144 lb per sq ft
1 psi = 70.3 g per sq cm
1 cm Hg (at 20° C) = 5.34 in. water (at 20° C)
1 cm Hg (at 20° C) = 13.57 cm water (at 20° C)
1 cm Hg (at 20° C) = 27.75 lb per sq ft
1 lb per sq ft = 0.488 g per sq cm
1 ton per sq ft = 0.976 kg per sq cm
1 ton per sq ft = 13.9 psi

Temperature

$T \text{ °C} = \frac{5}{9}(T \text{ °F} - 32°)$
$T \text{ °F} = \frac{9}{5}T \text{ °C} + 32°$
$T \text{ °K} = T \text{ °C} + 273.18°$

TABLE A–2. SPECIFIC GRAVITY OF WATER *

°C	0	1	2	3	4	5	6	7	8	9
0	0.9999	0.9999	1.0000	1.0000	1.0000	1.0000	1.0000	0.9999	0.9999	0.9998
10	0.9997	0.9996	0.9995	0.9994	0.9993	0.9991	0.9990	0.9988	0.9986	0.9984
20	0.9982	0.9980	0.9978	0.9976	0.9973	0.9971	0.9968	0.9965	0.9963	0.9960
30	0.9957	0.9954	0.9951	0.9947	0.9944	0.9941	0.9937	0.9934	0.9930	0.9926
40	0.9922	0.9919	0.9915	0.9911	0.9907	0.9902	0.9898	0.9894	0.9890	0.9885
50	0.9881	0.9876	0.9872	0.9867	0.9862	0.9857	0.9852	0.9848	0.9842	0.9838
60	0.9832	0.9827	0.9822	0.9817	0.9811	0.9806	0.9800	0.9795	0.9789	0.9784
70	0.9778	0.9772	0.9767	0.9761	0.9755	0.9749	0.9743	0.9737	0.9731	0.9724
80	0.9718	0.9712	0.9706	0.9699	0.9693	0.9686	0.9680	0.9673	0.9667	0.9660
90	0.9653	0.9647	0.9640	0.9633	0.9626	0.9619	0.9612	0.9605	0.9598	0.9591

* Also the density or unit weight of water in grams per milliliter.
From *International Critical Tables*, Vol. III, McGraw-Hill Book Co., 1928.

TABLE A–3. VISCOSITY OF WATER

(Values are in millipoises)

°C	0	1	2	3	4	5	6	7	8	9
0	17.94	17.32	16.74	16.19	15.68	15.19	14.73	14.29	13.87	13.48
10	13.10	12.74	12.39	12.06	11.75	11.45	11.16	10.88	10.60	10.34
20	10.09	9.84	9.61	9.38	9.16	8.95	8.75	8.55	8.36	8.18
30	8.00	7.83	7.67	7.51	7.36	7.31	7.06	6.92	6.79	6.66
40	6.54	6.42	6.30	6.18	6.08	5.97	5.87	5.77	5.68	5.58
50	5.29	5.40	5.32	5.24	5.15	5.07	4.99	4.92	4.84	4.77
60	4.70	4.63	4.56	4.50	4.43	4.37	4.31	4.24	4.19	4.13
70	4.07	4.02	3.96	3.91	3.86	3.81	3.76	3.71	3.66	3.62
80	3.57	3.53	3.48	3.44	3.40	3.36	3.32	3.28	3.24	3.20
90	3.17	3.13	3.10	3.06	3.03	2.99	2.96	2.93	2.90	2.87
100	2.84	2.82	2.79	2.76	2.73	2.70	2.67	2.64	2.62	2.59

1 dyne sec per sq cm = 1 poise
1 gram sec per sq cm = 980.7 poises
1 pound sec per sq ft = 478.69 poises
1 poise = 1000 millipoises

Data from *International Critical Tables*, Vol. V, McGraw-Hill Book Co., 1929.

PROVING RINGS

A proving ring consists of a ring, usually made of a high-carbon, heat-treated steel, with an extensometer mounted in such a way that it records radial deflections (see Figs. A–1 and A–3). With the relationship between load and deflection known, the applied load to cause any measured deflection can be determined. Because proving rings are an accurate and convenient means of measuring applied loads, they are widely used in both laboratory and in field soil testing apparatus.

Proving rings can be designed by the use of formulas [1] for the deflection and maximum fiber stress in radially loaded thin rings, and then machined to the designed thickness. Satisfactory rings have been made by grinding ball bearing races to desired thicknesses. The two rings in Fig. A–1 were made from ball bearing races; the track for the ball bearing is still visible in the ring on the right.

A ring is calibrated by applying known loads to it and measuring the resulting deflection. A typical calibration curve is shown in Fig. A–2. For this ring

[1] $\Delta = 0.149 \dfrac{Pr^3}{EI}$; $f_{max} = 0.318 \dfrac{Pry}{I}$

in which Δ = radial deflection at point where P is applied,
$\quad P$ = applied load,
$\quad r$ = radius of ring,
$\quad E$ = modulus of elasticity of ring,
$\quad I$ = moment of inertia of ring,
$\quad f_{max}$ = maximum fiber stress,
$\quad y$ = distance from neutral axis to point where f_{max} exists.

the ratio of load to deflection, or "calibration factor," is 0.5 lb per division, where one dial division is 0.0001 in. Since the load-deflection function is not exactly linear, there is a load correction for each deflection which must be added to the product of the calibration factor and deflection. For the ring whose curve is shown in Fig. A–2, the applied load, P, is found from

P = 0.5 lb per division

$\times$ number of division + correction

In most routine testing, the correction for this ring can be neglected.

To obtain greater accuracy in the low load range of tests which will later involve larger loads, and to prevent the necessity of changing rings for different tests, double proving rings have been developed (see Fig. A–3). When the more sensitive outer ring has reached its capacity, the thicker inside ring is brought into action. By this means advantage is taken of the high sensitivity of the outer ring without danger of overloading it.

Figure A–2 illustrates one of the undesirable features of the proving ring, namely, the apparent hysteresis effects. Loading and unloading the ring to about 50% capacity a few times before using will reduce these effects. This procedure, however, cannot always be followed. For example, it cannot be done prior to shear in a Q_c test. The proving frame (see Fig. A–4) does not exhibit the apparent hysteresis effects to the extent the ring does. The connection of the frame to the loading bar is a sounder one than

FIGURE A–1. Proving rings.

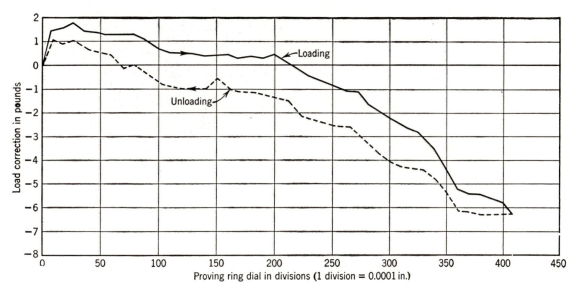

FIGURE A–2. Calibration curve for proving ring 12.

Capacity = 200 lb

Load in pounds = 0.5 dial in divisions + correction

that on the ring. Therefore, any contribution of the connection to the hysteresis is reduced. Although any looseness in the connection between the ring and its support can be removed by welding, it makes the stress-strain curve of the ring farther from a straight line.

The Amsler Company in Switzerland manufactures a dynamometer for measuring tensile or compressive forces. A dynamometer is an elongated elastic loop cut out of a solid block of hardened steel. This type of loop appears to give very precise measurements of load.

FIGURE A–3. Double proving ring. (Courtesy of Soil Testing Services, Inc., Chicago, Ill.)

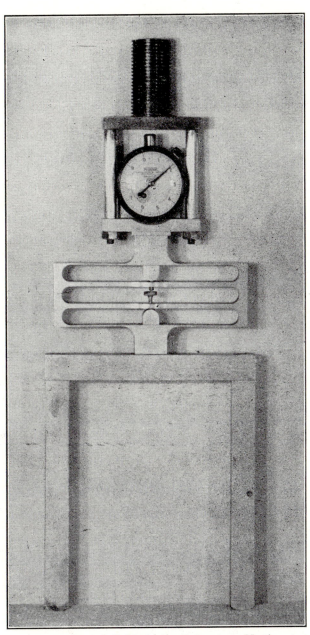

FIGURE A-4. Proving frame. (Courtesy of Soil Mechanics Laboratory, Northwestern University, Evanston, Ill.)

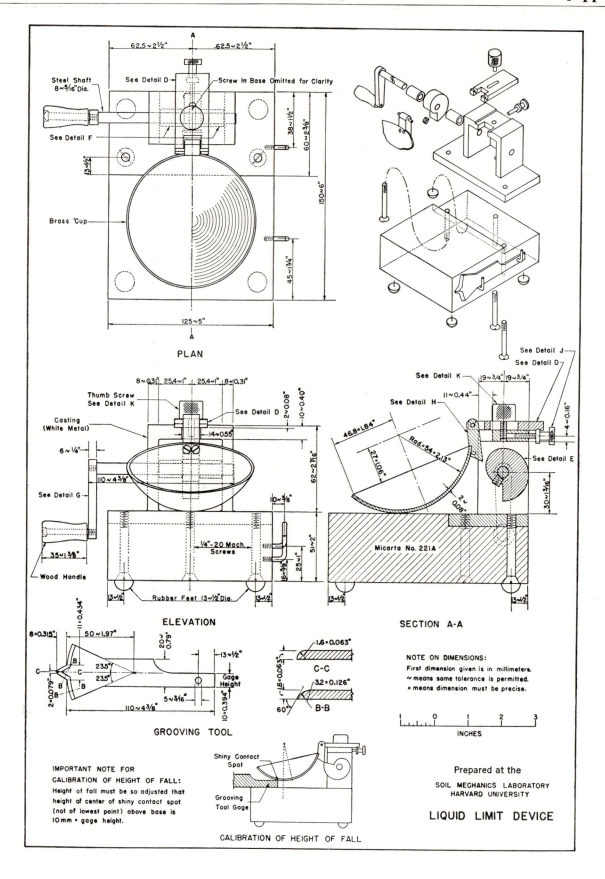

PLAN

ELEVATION

SECTION A-A

GROOVING TOOL

NOTE ON DIMENSIONS:
First dimension given is in millimeters.
~ means some tolerance is permitted.
= means dimension must be precise.

INCHES

IMPORTANT NOTE FOR
CALIBRATION OF HEIGHT OF FALL:
Height of fall must be so adjusted that
height of center of shiny contact spot
(not of lowest point) above base is
10 mm = gage height.

CALIBRATION OF HEIGHT OF FALL

Prepared at the
SOIL MECHANICS LABORATORY
HARVARD UNIVERSITY

LIQUID LIMIT DEVICE

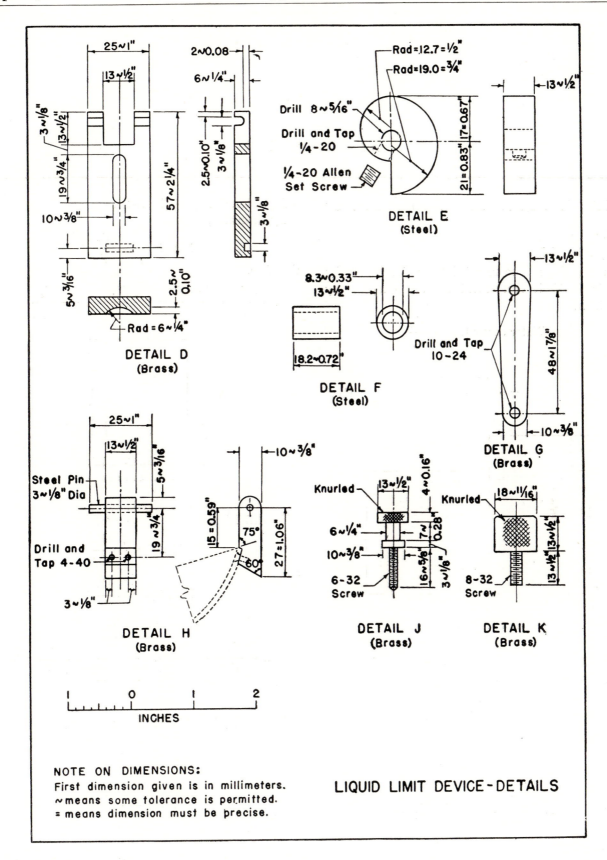

DETAIL D
(Brass)

DETAIL E
(Steel)

DETAIL F
(Steel)

DETAIL G
(Brass)

DETAIL H
(Brass)

DETAIL J
(Brass)

DETAIL K
(Brass)

INCHES

NOTE ON DIMENSIONS:
First dimension given is in millimeters.
~ means some tolerance is permitted.
= means dimension must be precise.

LIQUID LIMIT DEVICE-DETAILS

DERIVATION OF EQUATIONS[1]

Chapter I

Equation I–1.

In Fig. B–1a is represented the wet soil sample; in Fig. B–1b is represented the dry sample. By definition,

$$\text{Water content, } w = \frac{W_w}{W_s}$$

or

$$w = \frac{W_1 - W_2}{W_2 - W_c} \qquad (\text{I–1})$$

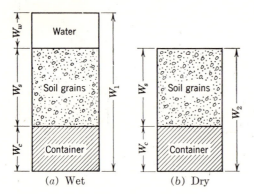

FIGURE B–1. Soil element.

Equation I–2 (see Fig. B–2).
By definition,

$$\text{Void ratio, } e = \frac{V_v}{V_s}$$

or

$$e = \frac{V - V_s}{V_s} = \frac{V}{V_s} - 1$$

$$= \frac{V}{W_s/G\gamma_w} - 1 = \frac{G\gamma_w V}{W_s} - 1 \qquad (\text{I–2})$$

[1] The nomenclature used in the following derivations is shown in the figures and is listed in Nomenclature, page ix.

Equation I–3 (see Fig. B–2).
By definition,

$$\text{Porosity, } n = \frac{V_v}{V}$$

therefore,

$$n = \frac{V_v}{V} = \frac{V - V_s}{V} = 1 - \frac{V_s}{V} = 1 - \frac{W_s}{G\gamma_w V} \qquad (\text{I–3})$$

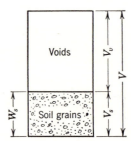

FIGURE B–2. Soil element.

Equation I–3a (see Fig. B–2).
By definition,

$$\text{Void ratio, } e = \frac{V_v}{V_s}$$

and

$$\text{Porosity, } n = \frac{V_v}{V}$$

therefore,

$$n = \frac{V_v}{V} = \frac{eV_s}{V_s + V_v} = \frac{e\dfrac{V_s}{V_s}}{\dfrac{V_s}{V_s} + \dfrac{V_v}{V_s}} = \frac{e}{1 + e} \qquad (\text{I–3}a)$$

Equation I–4 (see Fig. B–3).
By definition,

$$\text{Degree of saturation, } S = \frac{V_w}{V_v}$$

But

$$V_w = \frac{W_w}{\gamma_w}$$

therefore,

$$S = \frac{W_w}{\gamma_w V_v} \qquad (\text{I–4})$$

Equation I–5 (see Fig. B–3).

By definition,

$$w = \frac{W_w}{W_s} \; ; \quad e = \frac{V_v}{V_s} \; ; \quad S = \frac{V_w}{V_v} \; ; \quad G = \frac{\gamma_s}{\gamma_w}$$

Substituting in Eq. I–4, we have

$$S = \frac{w W_s}{\dfrac{\gamma_s}{G} e V_s} = \frac{wG}{e}$$

or

$$Se = Gw \qquad (I–5)$$

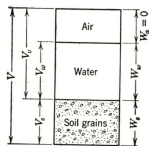

FIGURE B–3. Soil element.

Chapter II

Equation II–1.

Let W_2 = weight of pycnometer bottle (see Fig. II–1) plus water at temperature T,

W_B = weight of clean, dry bottle filled with air,

V_B = calibrated volume of bottle at temperature T_c,

V_{BT} = volume of bottle at temperature T,

T = temperature in degrees centigrade at which W_2 is desired,

T_c = calibration temperature of bottle (usually 20° C),

$\epsilon = 0.10 \times 10^{-4}/°C$,

γ_T = unit weight of distilled water at T,

γ_a = unit weight of air at T and atmospheric pressure (the average value of γ_a accurate enough for use in this test is 0.0012 g/cc).

From the foregoing definitions, it follows that

$$W_2 = W_B + V_{BT}\gamma_T - V_{BT}\gamma_a$$

or

$$W_2 = W_B + V_{BT}(\gamma_T - \gamma_a)$$

$$= W_B + [V_B + V_B(\Delta T \cdot \epsilon)](\gamma_T - \gamma_a)$$

or

$$W_2 = W_B + V_B(1 + \Delta T \cdot \epsilon)(\gamma_T - \gamma_a) \qquad (II–1)$$

Equation II–2.

By definition,

Specific gravity of solids, G_s

$$= \frac{\text{unit weight of soil, } \gamma_s}{\text{unit weight of water at 4° C, } \gamma_{w\,4°C}}$$

or

$$G_s = \frac{W_s/V_s}{\gamma_{w\,4°C}}$$

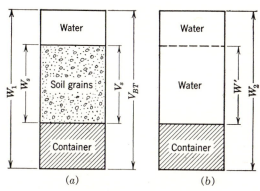

FIGURE B–4. Soil element.

and from Fig. B–4, where W' is the weight of water having a volume equal to that of the soil,

$$V_s = \text{the volume of water displaced} = \frac{W'}{\gamma_T}$$

Also

$$W' = W_2 + W_s - W_1$$

therefore,

$$G_s = \frac{W_s/V_s}{\gamma_{w\,4°C}} = \frac{W_s/(W'/\gamma_T)}{\gamma_{w\,4°C}}$$

$$= \frac{W_s\gamma_T}{(W_s - W_1 + W_2)\gamma_{w\,4°C}}$$

$$= \frac{W_s G_T}{W_s - W_1 + W_2} \qquad (II–2)$$

Chapter III

Equation III–1 (see Fig. B–5).

By definition, the shrinkage limit, w_s, is the water content of a soil specimen which is completely satu-

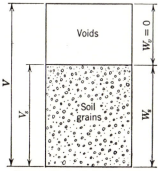

FIGURE B–5. Soil element.

$$V - V_S = V - (W_S/\gamma_S)$$

rated and at its minimum volume obtained by drying. V is the volume of soil when it is at the shrinkage limit. Therefore,

$$w_s = \frac{W_w}{W_s} = \frac{(V - V_s)\gamma_w}{W_s} = \frac{V - (W_s/\gamma_s)}{W_s}\gamma_w$$

or

$$w_s = \frac{\gamma_w V}{W_s} - \frac{\gamma_w}{\gamma_s} = \frac{\gamma_w V}{W_s} - \frac{G_T}{G_s} \qquad \text{(III–1)}$$

Chapter IV

Hydrometer Calibration.

Casagrande has shown that in a liquid of variable specific gravity a hydrometer reads the specific gravity

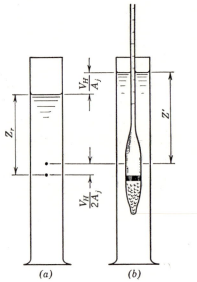

FIGURE B–6. Graduate and hydrometer.

at approximately the depth in the liquid where the center of volume of the hydrometer floats (IV–1). Since it is this depth that must be used in Eq. IV–1, it is desirable to have a curve which gives the depth from the surface of the suspension to the center of volume of the hydrometer for any hydrometer reading. Such a curve, known as a calibration curve, is essentially a straight line for a symmetrical bulb hydrometer such as is shown in Figs. IV–2a and B–6b. To calibrate such a hydrometer, first locate the mid-length of the bulb, which is very close to the center of volume. Next measure the distance from a graduation mark on each end of the stem to the center of the bulb; because the curve is essentially straight, only two such measurements are needed. Plot a curve of hydrometer reading against depth, as curve A, Fig. B–7.

For readings within the first 2 minutes of a hydrometer test, the hydrometer is left in the suspension, but for the remaining readings it is inserted before and removed after each reading. When the hy-

drometer is inserted into the suspension, the surface of the suspension rises and, therefore, an immersion correction must be applied. The immersion correction is shown in Fig. B–6; the distance Z' must be corrected to Z_r. The relationship between Z' and Z_r is

$$Z_r = Z' - \frac{V_H}{2A_j}$$

where V_H = volume of the hydrometer,[2]
A_j = the area of the graduated jar.

The expression above is reasoned as follows. When the hydrometer is inserted, the surface of the suspension rises a distance equal to V_H/A_j; the center of the hydrometer is, in effect, raised an amount $V_H/2A_j$ because of the displacement of the lower half of the hydrometer; the combination of these two effects gives $Z_r = Z' - (V_H/2A_j)$.

The cross-sectional area of the jar, A_j, is obtained by dividing the volume between two calibration marks by the distance between the same two marks. The volume [3] of the hydrometer can be obtained by im-

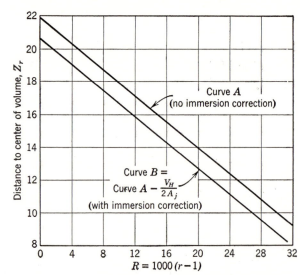

FIGURE B–7. Calibration curve for hydrometer 7365.

mersing it in a graduate of water and noting the increase of volume as read on the graduate. Another method of determining the hydrometer volume is to calculate it from its weight and unit weight. The unit weight can be taken as equal to the lowest reading on

---

[2] Actually V_H is the volume of that part of the hydrometer which is immersed; therefore, V_H depends on the hydrometer reading. The value of V_H is usually taken as constant since the amount which it varies is insignificant.

[3] A detailed method of measuring V_H is given in "Procedures for Testing Soils," American Society for Testing Materials, July, 1950.

the specific gravity scale multiplied by the unit weight of water.

After determining the immersion correction, $V_H/2A_j$, plot a curve, such as curve B, Fig. B–7, by subtracting the correction from curve A. When using Eq. IV–1, read curve A for the observations in the first two minutes and read curve B for the other observations. Allowance should be made for the rise of meniscus on the hydrometer stem before entering the hydrometer calibration curve. This allowance is made by numerically adding the height of meniscus rise, as determined in pure water at the temperature of the test, to the hydrometer reading.

Equation IV–1.

The equation comes from Stokes' law for the terminal velocity, v, of a freely falling sphere, which is,

$$v = \frac{\gamma_s - \gamma_w}{18\mu} D^2$$

where D = the sphere diameter

or

$$D = \sqrt{\frac{18\mu}{\gamma_s - \gamma_w}} \sqrt{v}$$

The velocity of particle fall is also equal to the distance of fall, Z_r, divided by the elapsed time, t, or

$$v = \frac{Z_r}{t}$$

Therefore,

$$D = \sqrt{\frac{18\mu}{\gamma_s - \gamma_w}} \sqrt{\frac{Z_r}{t}} \qquad \text{(IV–1)}$$

Equation IV–2.

If a soil-water suspension is completely mixed, the unit weight, γ_i, of the suspension is

$$\gamma_i = \frac{W_s}{V} + \left(\gamma_w - \frac{W_s}{GV}\right)$$

where W_s = the weight of soil,
           V = the volume of container (i.e., graduate),

or

$$\gamma_i = \gamma_w + \frac{G-1}{G}\frac{W_s}{V}$$

If the suspension is allowed to settle, at any time, t, and down a distance, Z, the distribution of all grains finer than D, which is the diameter of the smallest grain which could have fallen Z in t, remains unchanged. Let N_1 be the ratio of the weight of soil

grains smaller than D to the total weight of the soil. The weight of solids per unit of volume at depth Z and time t is N_1W_s/V, and the suspension unit weight, γ, is

$$\gamma = \gamma_w + \frac{G-1}{G}\frac{N_1W_s}{V}$$

or

$$N_1 = \frac{G}{G-1}\frac{V}{W_s}(\gamma - \gamma_w)$$

or, in percentage,[4]

$$N = \frac{G}{G-1}\frac{V}{W_s}\gamma_c(r - r_w) \times 100\% \qquad \text{(IV–2)}$$

It should be noted that in Eq. IV–2 the difference of hydrometer readings in the suspension and distilled water, or $r - r_w$, cancels any effect of temperature or meniscus rise on the stem.

Equation IV–3 (see Fig. B–8).

In Fig. B–8,

W_s = total soil weight used in combined analysis
W_1 = weight of soil finer than No. 200 sieve
W_2 = weight of soil used in hydrometer analysis

When W_2 is used for W_s in Eq. IV–2 to get N, then N is the percentage finer based on all the material pass-

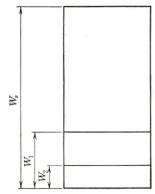

FIGURE B–8. Weights in grain size analysis.

ing the No. 200 sieve. This is true if W_2 is a representative sample of W_1. Therefore, to reduce N to a basis of the entire soil, W_s,

$$N' = N \cdot \frac{W_1}{W_s} = N \cdot \% \text{ finer than No. 200 sieve} \qquad \text{(IV–3)}$$

where N' = percentage finer based on entire soil sample, W_s.

---

[4] This derivation is abbreviated from that in Taylor's *Fundamentals of Soil Mechanics*.

Chapter V

Equation V–1 (see Fig. B–9).

By definition, the dry density,[5] $\gamma_d = \dfrac{W_s}{V}$

Also, by definition,

$$w = \frac{W_w}{W_s}$$

Therefore,

$$w = \frac{W - W_s}{W_s} \; ; \quad W_s = \frac{W}{1 + w}$$

Substituting, we get

$$\gamma_d = \frac{W_s}{V} = \frac{W}{V(1 + w)} \qquad \text{(V–1)}$$

For the original Proctor mold of $\frac{1}{30}$ cu ft volume,

$$\gamma_d = \frac{30W}{1 + w} \qquad \text{(V–1a)}$$

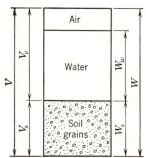

FIGURE B–9. Soil element.

Equation V–2.
From Fig. B–9,

$$\gamma_d = \frac{W_s}{V} = \frac{W_s}{V_s + V_v} = \frac{W_s/V_s}{\dfrac{V_s}{V_s} + \dfrac{V_v}{V_s}} = \frac{\gamma_s}{1 + e} = \frac{G\gamma_w}{1 + e}$$

Also,

The void ratio, $e = \dfrac{V_v}{V_s} = \dfrac{W_w}{W_w} \cdot \dfrac{W_s}{W_s} \cdot \dfrac{V_w}{V_w} \cdot \dfrac{V_v}{V_s}$

$$= \frac{\dfrac{W_w}{W_s} \cdot \dfrac{W_s}{V_s} \cdot \dfrac{V_w}{W_w}}{V_w/V_v} = \frac{wG}{S}$$

Therefore,

$$\gamma_d = \frac{G\gamma_w}{1 + (wG/S)} \qquad \text{(V–2)}$$

[5] Dry density is a misnomer. It is a unit weight, not a density: the volume involved is that of the wet soil mass.

Equation V–3.
From Fig. B–9,

$$\gamma_d = \frac{W_s}{V} = \frac{W_s}{V_s + V_v} = \frac{W_s/V_s}{\dfrac{V_s}{V_s} + \dfrac{V_v}{V_s}}$$

$$= \frac{\gamma_s}{1 + e} = \frac{G}{1 + e}\gamma_w \qquad \text{(V–3)}$$

Chapter VI

Equation VI–1 (see Fig. VI–2).
In the standpipe,

The rate of flow, $q = \text{velocity} \times \text{area} = -\dfrac{dh}{dt}a$

(the minus is to take care of the fact that h decreases as t increases)

In the permeameter, by Darcy's law,

The rate of flow, $q = kiA = k\dfrac{h}{L}A$

Equating the two expressions for q gives

$$k\frac{h}{L}A = -\frac{dh}{dt}a$$

or

$$k = 2.3\frac{aL}{A(t_1 - t_0)}\log_{10}\frac{h_0}{h_1} \qquad \text{(VI–1)}$$

Equation VI–2 (see Fig. VI–3).
From Darcy's law, the rate of flow, q, through the soil is

$$q = kiA = k\frac{h}{L}A$$

Also the rate of flow, q, is

$$q = \frac{Q}{t}$$

where Q = the quantity of water flowing in the graduate in time, t.

Equating the two expressions for q gives

$$k\frac{h}{L}A = \frac{Q}{t}$$

or

$$k = \frac{QL}{thA} \qquad \text{(VI–2)}$$

Equation VI–3.
By an analogy of flow through soils to flow through capillary tube, and by using Poiseuille's law for flow

through capillary tubes, the following equation for soil flow has been derived,[6]

$$k = D_s^2 \frac{\gamma_w}{u} \frac{e^3}{1+e} C$$

In the equation above all terms except the fluid viscosity, μ, are insensitive to temperature. Therefore, for temperature changes, we can write,

$$k_{20°C} = k \frac{\mu_T}{\mu_{20°C}} \tag{VI–3}$$

Chapter VIII
Equation VIII–1.

By Darcy's law, the rate of flow, q, in the wetted soil of Fig. B–10 is

$$q = kiA = k_s \frac{h_0 + h_c'}{x} A$$

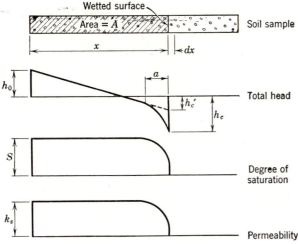

FIGURE B–10. Horizontal capillary flow.

Assuming that a and h_c' are constant,

$$q = \frac{dx}{dt} SnA$$

where n is the porosity of the soil. Therefore,

$$\frac{dx}{dt} SnA = k_s \frac{h_0 + h_c'}{x} A$$

Solution of the above equation gives

$$\frac{\Delta(x^2)}{\Delta t} = \frac{2k_s}{Sn}(h_0 + h_c') \tag{VIII–1}$$

In the derivation above it should be noted that h_c is the true capillary head and h_c' is an effective head required to give a constant gradient for the length x. Thus h_c' is no fundamental soil property, but depends

[6] Taylor, *Fundamentals of Soil Mechanics*, Art. 6·5.

to a minor degree on the value of x. For the normal laboratory test, h_c' and a can be taken as constant.

Chapter IX
Equation IX–1 (see Fig. B–11).

By definition,

$$W_s = V_s G \gamma_w$$

Also

$$V_s = 2H_0 A$$

where $2H_0$ = the height of solids, and
A = the specimen area.

Therefore,

$$W_s = 2H_0 A G \gamma_w$$

or

$$2H_0 = \frac{W_s}{G\gamma_w A} \tag{IX–1}$$

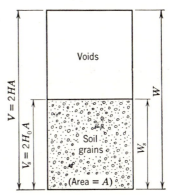

FIGURE B–11. Soil element.

Equation IX–3.

Equations IX–3a and IX–3b follow from the definition of primary compression ratio and the use of $d_s - d_{100} = \frac{10}{9}(d_s - d_{90})$, where the terms are defined on page 83.

Equation IX–4 (see Fig. B–12).

By definition, the compression index, C_c, is the slope of the log pressure, p, versus void ratio, e, curve or

$$C_c = \frac{-de}{d(\log_{10} p)} \tag{IX–4}[7]$$

Equation IX–5 (see Fig. B–12).

From Eq. IX–4,

$$de = -C_c d(\log_{10} p)$$

Differentiation gives

$$d(\log_{10} p) = \frac{\log_{10} e}{p} dp = \frac{0.435 dp}{p}$$

[7] The minus sign is put in to take care of the fact that e decreases as p increases.

By definition,

$$a_v = \frac{-de}{dp} \, [7]$$

or

$$a_v = \frac{0.435 C_c}{p} \qquad \text{(IX–5)}$$

where p is usually taken as the average pressure for the increment from p_1 to p_2 or

$$p = \frac{p_1 + p_2}{2}$$

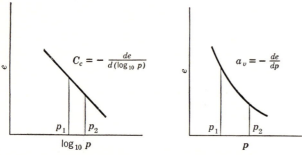

FIGURE B–12. Compression curves.

Equation IX–6.

In the derivation of the Terzaghi theory of consolidation (IX–10) the coefficient of consolidation, c_v, is defined as

$$c_v = \frac{k(1 + e)}{a_v \gamma_w}$$

Rearranging the above equation gives

$$k = \frac{c_v a_v \gamma_w}{1 + e} \qquad \text{(IX–6)}$$

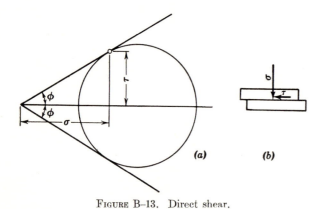

FIGURE B–13. Direct shear.

Chapter X
Equation X–1.

In Fig. B–13a is the Mohr circle for the state of stress at failure in the direct shear test shown in Fig. B–13b. It can be seen in this figure that

$$\tan \phi = \frac{\tau}{\sigma}$$

or

$$\phi = \tan^{-1} \frac{\tau}{\sigma} \qquad \text{(X–1)}$$

Chapter XI
Equation XI–3.

If it is assumed that the volume of specimen remains constant during shear, the volume, V, at any time is equal to the initial volume V_0. Therefore,

$$V = V_0 = AL = A_0 L_0$$

in which A and L are the specimen area and length at any given strain. By definition,

$$L = L_0 - \Delta L$$

or

$$L = L_0 - \epsilon L_0 = L_0(1 - \epsilon)$$

where ϵ is the axial strain and

$$A = \frac{A_0 L_0}{L} = \frac{A_0 L_0}{L_0(1 - \epsilon)} = \frac{A_0}{1 - \epsilon} \qquad \text{(XI–3)}$$

Equation XI–3 is, of course, in error if the assumptions upon which it is based are not fulfilled. Although it was assumed that the specimen volume was constant during shear, a look at the Numerical Example, page 108, shows that this is not true. For example, at a strain of 15.1%, the area obtained by assuming a constant volume was 7.15 sq in. Actually the volume at 15.1% strain was 35.7 cc greater than the initial volume. If the correct volume is used in computing the area, a value of 7.53 sq in. is obtained.

Equation XI–3 also assumes uniform strain during compression. Occasionally in compressing a cohesionless soil and often in compressing a cohesive soil, the strains are far from uniform. For example, it is possible that nearly all the compression and shear occur in only half the length, thereby making the effective strains twice the apparent values. If this were to occur in the Numerical Example, page 108, the area when the apparent strain was 15.1 would be 8.65 sq. in., which is 17.5% greater than the 7.15 sq in. area.

Equation XIII–5 (see page 129) takes care of changes in volume and thus should be used instead of Eq. XI–3, if greater accuracy is desired. The equation given in footnote 20 on page 129 approximately takes care of the nonuniform strains.

Fidler (XI–4) ran a series of triaxial tests on sand in which he measured the maximum diameter, d_{max}, during shear by means of a tape. He then figured the area of the elipse whose major diameter was d_{max} and whose minor diameter was the initial diameter of the specimen. Such an area closely approximates the horizontal projection of the plane of maximum obliquity, which is the effective area over which the vertical stress is applied. For three such tests, he obtained areas which were within 5% of those calculated by Eq. XI–3.

For routine tests in which volume changes are small and strains appear to be reasonably uniform, Eq. XI–3 is probably acceptable and is normally used.

Equation XI–5.

Since $\bar{\sigma}_1 = p + \bar{\sigma}_3$,

$$\frac{\bar{\sigma}_1}{\bar{\sigma}_3} = \frac{p + \bar{\sigma}_3}{\bar{\sigma}_3} \qquad (XI–5)$$

Equation XI–6.

In Fig. B–14 is the Mohr circle for the failure stresses in the triaxial test. By geometry, from the figure,

$$\sin\phi = \frac{BC}{AB} = \frac{\dfrac{\bar{\sigma}_1 - \bar{\sigma}_3}{2}}{\dfrac{\bar{\sigma}_1 + \bar{\sigma}_3}{2}} = \frac{\dfrac{\bar{\sigma}_1}{\bar{\sigma}_3} - 1}{\dfrac{\bar{\sigma}_1}{\bar{\sigma}_3} + 1}$$

Therefore,

$$\phi = \sin^{-1}\frac{(\bar{\sigma}_1/\bar{\sigma}_3) - 1}{(\bar{\sigma}_1/\bar{\sigma}_3) + 1} \qquad (XI–6)$$

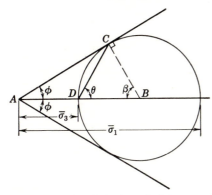

FIGURE B–14. Mohr circle for triaxial compression.

Equation XI–7.

Use of geometry in Fig. B–14 gives

$$2\theta = 180° - \beta$$

and

$$\phi + \beta = 90°$$

where θ = the angle between the failure plane DC and the major principal plane DB.

Substitution gives

$$2\theta = 180° - (90° - \phi)$$

or

$$\theta = 45° + \frac{\phi}{2} \qquad (XI–7)$$

Chapter XII
Equation XII–1.

In Fig. B–15 is shown the Mohr circle for an unconfined compression test. Since the pore water pres-

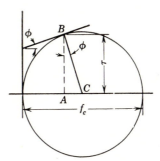

FIGURE B–15. Mohr circle for unconfined compression.

sures are not known, only the applied stress can be plotted. The shear strength, AB, is equal to BC $\cos\phi$, or

$$\tau = BC \cos\phi = \frac{f_c}{2} \cos\phi$$

The friction angle, ϕ, is normally about 30°; therefore, from the expression above the shear strength would be approximately $0.43f_c$.

More elaborate tests in which the friction angle has been measured have shown that the shear strength of a clay is nearer 0.5 times the compressive strength as measured by the unconfined compression test. The results mentioned above indicate that there is something inherent in the unconfined test which gives a value of compressive strength lower than that given by other types of tests. The cause of this lower strength is probably due to the severe condition of no lateral support to the specimen. In view of the foregoing, the value of shear strength is commonly taken equal to one-half the unconfined compressive strength, or

$$\tau = \frac{f_c}{2} \qquad (XII–3)$$

Chapter XIII
Equation XIII–1.

Equation XIII–1 is derived from the definition of strain, with the specimen length just prior to shear being used as the "initial" specimen length.

Equation XIII–2.

Adding the deviator stress, p, to the chamber pressure, which is the minor principal stress, σ_3, gives the major principal stress, σ_1, or

Since
$$\sigma_1 = p + \sigma_3 \qquad \text{or} \qquad p = \sigma_1 - \sigma_3$$

$$\bar{\sigma}_1 = \sigma_1 - u \qquad \text{and} \qquad \bar{\sigma}_3 = \sigma_3 - u$$

where u is the pore water pressure, it follows that

$$p = \sigma_1 - u - \sigma_3 + u = (\sigma_1 - u) - (\sigma_3 - u)$$

or
$$p = \bar{\sigma}_1 - \bar{\sigma}_3 \qquad \text{(XIII–2)}$$

Equation XIII–3.

Equation XIII–3 is derived like Equation XI–3, where the average area after consolidation, A_c, is the average area just before shear.

Equation XIII–4.

The specimen volume after consolidation, V_c, is equal to the volume prior to consolidation, V_0, minus the volume change during consolidation, or

$$V_c = V_c - \Delta V_c$$

Also V_c is equal to the average area after consolidation, A_c, times the length after consolidation, L_c, or

$$V_c = A_c L_c$$

But
$$L_c = L_0 - \Delta L_c$$

where $L_0 =$ the initial length, and
$\Delta L_c =$ the change of length during consolidation.

Substitution gives

$$V_c = V_0 - \Delta V_c = A_c L_c = A_c(L_0 - \Delta L_c)$$

or
$$A_c = \frac{V_0 - \Delta V_c}{L_0 - \Delta L_c} \qquad \text{(XIII–4)}$$

Equation XIII–5.

The volume, V, at any time is

$$V = V_0 - \Delta V$$

where $V_0 =$ the initial volume, and
$\Delta V =$ the volume change.

Also
$$V = LA = (L_0 - \Delta L)A$$

where $L_0 =$ the initial length,
$\Delta L =$ the change in length, and
$A =$ the average area at any time.

Substitution gives

$$V = V_0 - \Delta V = (L_0 - \Delta L)A$$

or
$$A = \frac{V_0 - \Delta V}{L_0 - \Delta L} \qquad \text{(XIII–5)}$$

Chapter XIV
Equation XIV–1.

By definition, shear stress is equal to shear force divided by the area of the shear surface, or

$$\tau = \frac{F}{A} \qquad \text{(XIV–1)}$$

Index